48 Advances in Biochemical Engineering Biotechnology

Managing Editor: A. Fiechter

Bioprocess Design and Control

With contributions by
R. Aarts, M. Aynsley, J. E. Bailey, P. M. Doran,
I. J. Dunn, K. Friehs, E. Heinzle, A. Hofland,
K. B. Konstantinov, C. Di Massimo,
G. A. Montague, A. J. Morris, K. F. Reardon,
G. B. Ryhiner, T. Yoshida

With 52 Figures and 21 Tables

Springer-Verlag
Berlin Heidelberg GmbH

ISBN 978-3-662-14959-1 ISBN 978-3-540-47517-0 (eBook)
DOI 10.1007/978-3-540-47517-0

Originally published by Springer-Verlag Berlin Heidelberg New York in 1993
Softcover reprint of the hardcover 1st edition 1993

Library of Congress Catalog Card Number 72-152360

Typesetting: Macmillan India Ltd., Bangalore-25

02/3020 - 5 4 3 2 1 0 Printed on acid-free paper

Attention all "Enzyme Handbook" Users:

A file with the complete volume indexes Vols. 1 through 5 in delimited ASCII format is available for downloading at no charge from the Springer EARN mailbox. Delimited ASCII format can be imported into most databanks.

The file has been compressed using the popular shareware program "PKZIP" (Trademark of PKware INc., PKZIP is available from most BBS and shareware distributors).

This file distributed without any expressed or implied warranty.

To receive this file send an e-mail message to:
SVSERV@DHDSPRI6.BITNET.
The message must be: "GET/ENZHB/ENZ_HB.ZIP".

SPSERV is an automatic data distribution system. It responds to your message. The following commands are available:

HELP	returns a detailed instruction set for the use of SVSERV,
DIR (*name*)	returns a list of files available in the directory "name",
INDEX (*name*)	same as "DIR"
CD <*name*>	changes to directory "name",
SEND <*filename*>	invokes a message with the file "filename"
GET <*filename*>	same as "SEND".

Table of Contents

Artificial Intelligence and the Supervision of Bioprocesses (Real-Time Knowledge-Based Systems and Neural Networks)

M. Aynsley, A. Hofland, A.J. Morris, G.A. Montague and C. Di Massimo
Department of Chemical and Process Engineering, University of Newcastle, Newcastle upon Tyne, NE1 7RU, England

Dedicated to Prof. Karl Schügerl on the occasion of his 65th birthday

The ability to supervise and control a highly non-linear and time variant bioprocess is of considerable importance to the biotechnological industries which are continually striving to obtain higher yields and improved uniformity of production. Two AI methodologies aimed at contributing to the overall intelligent monitoring and control of bioprocess operations are discussed.

The development and application of a real-time knowledge-based system to provide supervisory control of fed-batch bioprocesses is reviewed. The system performs sensor validation, fault detection and diagnosis and incorporates relevant expertise and experience drawn from both bioprocess engineering and control engineering domains. A complementary approach, that of artificial neural networks is also addressed. The development of neural network modelling tools for use in bioprocess state estimation and inferential control are reviewed. An attractive characteristic of neural networks is that with the appropriate topology any non-linear functional relationship can be modelled, hence significantly reducing model-process mismatch. Results from industrial applications are presented.

Advances in Biochemical Engineering
Biotechnology, Vol. 48
Managing Editor: A. Fiechter

Abbreviations

AI	artificial intelligence
CER	carbon dioxide evolution rate
ICP	intelligent communication protocol
KBS	knowledge-based system
OUR	oxygen uptake rate
RQ	respiratory quotient
RTKBS	real-time knowledge-based system
SQL	Structural query language
TCS	Turnbull control system

1 Introduction

The requirement to operate biotechnological processes in a cost effective manner is becoming increasingly relevant as the bioprocessing industry develops and market competition increases. This situation is seen both in the more established product areas, such as penicillin production, as well as in more recently developed processes such as the production of protein. Historically, the most important means of achieving increased productivity from bioprocess plant has been through strain improvements and media developments. More recently, however, significant advances have been made in the area of bioprocess supervision and control. This alternative, complementary, route to process improvement has many potential benefits, although many problem areas are still to be overcome before the application of advanced supervision and control techniques becomes common place.

Typically, little use is made of the large amounts of bioprocess on-line and laboratory assay data after a batch is completed (Stephanopoulos and Tsiveriotis [1]). Recently, attempts have been made using knowledge-based systems (KBS) approaches to improve the quality of information presented to operators (Karim and Halme [2]) and increase the level of automatic process supervision (Cooney et al. [3]), although few industrial applications have yet been reported. On a plant-wide perspective, the scheduling of process operations is also traditionally manually undertaken by experienced personnel using a trial and error approach with a planning board. This also appears to be a potential area for the application of a KBS (Hofmeister et al. [4]). In the particular fed batch bioprocess described, the "out of sequence" preferential harvesting to maximize productivity, while ensuring that this would not result in lengthy down-times for other process units, could yield a very significant improvement in the economics of production.

Knowledge-based systems can be designed to cope with uncertainty and allow the coupling of quantitative information with the qualitative or symbolic expressions (in the form of heuristics) so as to reproduce the actions of an experienced process operator. They might therefore be used as an intelligent, on-line assistant to the process operator, or, with extra knowledge as a supervisor in running and maintaining bioprocess operations within optimal operating conditions. The main objective in this approach is to develop an expert supervisory system which uses bioprocess and control knowledge in the form of rules in conjunction with bioreactor state estimation (soft-sensing), parametric identification and process control algorithms running in real-time. In addition, provision needs to be made for process models and known bioprocess behavioural characteristics to be incorporated in order to enhance the overall supervisory control strategy. The complete system aims to perform process monitoring, process control, sensor validation, fault detection and diagnostic tasks and in addition, provide recovery advice so as to achieve continuous on-line optimisation.

An integrated system for improved supervisory control and scheduling of bioprocesses is proposed using a KBS approach. The system will comprise a bioprocess supervisory knowledge base and a plant scheduling knowledge base, both of which will be linked to an on-line relational database to provide a maximum level of integrity and intelligence in a real-time environment.

Collateral with the above approach is the on-line determination of the critical variables which govern process production (e.g. biomass concentration, growth rate etc.). Halme [5], and Montague et al. [6] have reviewed the measurement and estimation problems and techniques in bioprocess systems. Halme [5] indicates three ways of reducing the difficulties encountered:

- better sensors
- better sampling and automatic analysis systems
- on-line estimation of difficult to measure, or unmeasurable, variables.

Although all of these are attracting research effort, the techniques of on-line state and parameter estimation are receiving the greatest interest.

Whilst developments in biosensors are in some cases helping to contribute to the on-line determination of bioprocess parameters, they are by no means at the stage where general applicability has been achieved. Until this is the case the development of on-line estimation techniques will provide the major means by which on-line closed loop feedback control of critical bioprocess variables can be achieved. (The need to obtain information from process variables which are difficult to measure is not restricted to any one type of process. Problems can be identified in such diverse industries as for example the processing of food, chemicals, polymers, minerals, etc.)

Several different approaches can be adopted in the development of bioprocess estimation algorithms. The easiest, but potentially least accurate, way of obtaining an estimate of a "difficult to measure" primary process variable is to establish a correlation with a measurable secondary process variable, whilst ignoring measurement errors, noise, etc. A more robust approach might be to adopt a numerical estimation technique, either for estimating the parameters of a pre-defined model structure, usually of a generalised linear time series form, or directly to obtain an estimate of a difficult to measure primary process variable (state estimation or soft-sensor). An alternative approach is to exploit the nonlinear mechanistic structure of the bioprocess, if known, in the form of a state observer.

A relatively new development in artificial intelligence is the area of neural computing. Neural networks are dynamic systems composed of highly interconnected layers of "simple" neurone-like processing elements. In the field of process engineering it is their ability to capture process non-linearities which offers the potential benefits. Other useful characteristics are their ability to adjust dynamically to environmental and time-variant changes, to infer general rules from specific examples, and to recognize invariances from complex high-dimensional data. These properties provide neural networks with the potential to out-perform other "learning" techniques. Given a series of examples the network is able to establish the governing relationships in the training data. This ability can be exploited to aid the nonlinear modelling, control and optimisation of complex processes, as well as being used in existing predictive control methods.

Two industrial systems are used to demonstrate the application of the AI methodologies reviewed – the fed-batch production of penicillin G and a large scale mycelial process which operates in a continuous mode.

2 The Bioprocesses Studied

2.1 An Industrial Penicillin Process

The production of an established product such as penicillin is a highly competitive business. Even small savings that could be achieved through the application of improved supervision and control are important in the gaining of a market edge. The industrial production of penicillin is achieved by a fed-batch process in which two distinct operating regimes can be identified. In the early stages the process is operated to produce large quantities of biomass, predominantly utilising the substrate in the initial batched media. Towards the end of this phase the feed additions being made to the bioreactor are increased as the initial substrate becomes exhausted. During the second phase the substrate additions are maintained at a rate which keeps the substrate concentration at a low level. As a consequence of the low substrate concentration and the resulting low growth rate, penicillin is produced by the organism. In order to maximize the yield of penicillin it has been observed that the growth rate should be maintained above a pre-determined minimum value. The higher the growth rate is above this minimum constraint the lower the yield of penicillin.

The present operating regime of off-line analysis to determine the bioreactor conditions results in a conservative feeding strategy since samples are taken relatively infrequently. It is highly desirable therefore to gain some on-line insight as to bioprocess behaviour at a higher frequency so as to be able to operate closer to the constraint. To increase the frequency of information routinely available would require a move from off-line analysis to on-line measurement or estimation. Unfortunately, to date instrumentation is not available which provides the necessary data. It is for this reason that an observer has been developed to provide this information and utilise it within a control strategy in order to improve overall bioprocess operation.

2.2 An Industrial Mycelial Process

This particular bioprocess (*Fusarium graminearum*) is operated in a continuous mode, on a large scale, to produce biomass which on processing is marketed as a meat analogue. It is described in more detail in Edelman et al. [7]. Regulation of the biomass concentration is required in order to achieve the desired product quality. The present regulatory philosophy is based upon the use of off-line biomass assays, carried out every four hours. The results are available to the process operator one to three hours after sampling. Based upon the assay result, the dilution rate is adjusted in an attempt to compensate for process disturbances. The shortcoming of this control scheme is the low sampling frequency of biomass concentration in relation to the process dynamics. Furthermore, because the laboratory analysis of dry weight is subject to significant error, a biased averaging of the last three biomass assays is applied. Consequently, data up to 15 h old is utilized for feedback control of biomass. It is not therefore surprising that the control scheme results in excursions outside acceptable operating conditions. Carbon dioxide evolution rate measurements are available to the process operators almost continuously and have been shown to be a good indicator of the level of biomass present in the bioreactor. However, the relationship between biomass concentration and carbon dioxide evolution rate has not been quantitatively analysed.

3 Real-Time Knowledge-Based Supervision of Bioprocesses

Real-Time Knowledge-Based Systems (RTKBS) can provide novel solutions to many problems encountered in the operation of time critical processes. Chantler [8], Stephanopoulos and Stephanopoulos [9] and Moore et al. [10] have discussed the basic requirements of RTKBS which are now becoming well established in a wide range of application domains. It is only very recently, however, that reports have appeared on the application of expert systems in the field of bioprocess control [2, 3, 11].

3.1 The Real-Time Knowledge-Based System Shell

The G2 Real-Time Expert System [12] provides a flexible framework within which to construct a knowledge base, but with much more powerful features than those usually provided by a shell. The concept of real-time is represented within the system in terms of update intervals which direct the supervisory software to collect values for variables at regular time intervals; validity intervals which specify the length of time the current value of a variable remains relevant, and rule scanning intervals which inform the inference engine how often to invoke any given rule. Currently valid data can be reasoned with directly or can be analysed alongside past data using statistical features. It is also possible to import functions written in "C" or Fortran which can be used to carry out complex numerical calculations. Rules are written in a very expressive English-like language and have an icon based graphical interface to create knowledge frames (i.e. an object-orientated approach). The system also supports arithmetic and symbolic expressions and has a built-in simulation facility which can be used in the "background" while the actual process is being controlled in real-time.

3.2 Bioprocess Supervision Control and Analysis (Bio-SCAN)

Bio-SCAN is a real-time knowledge-based system, developed using G2, which has been designed to provide a general purpose framework for the monitoring and supervisory control of a range of bioprocesses. Internally, the system incorporates knowledge of operating strategies (e.g. feeding regimes, process set-points) and expected behaviours to enable rapid detection of any process faults. When faults occur, the system attempts to diagnose the cause of the problem and will offer recovery advice to the process operators. In addition to knowledge encoded within Bio-SCAN, the utility of the system has been extended by interfacing to conventional algorithmic methods of analysing and controlling bioprocesses. A toolkit of on-line modules has been developed to incorporate a number of features into the overall control strategy such as simulation and prediction, statistical evaluation and bioreactor state estimation algorithms. An underlying philosophy in the development of Bio-SCAN has been to make the solutions as generic as possible (applicable across a broad spectrum of bioprocesses) and to provide the system with as much (high quality) information as possible. This led to the development of a single data depository (database) for on-line interrogation by Bio-SCAN. Further details about the development of the system can be found in Aynsley et al. [13].

System development begins with the introduction of objects necessary for the description of the process under study. The objects are assigned attributes (usually symbolic or quantitative variables) and the source of their values. Rules are written which inform Bio-SCAN about what to conclude or how to respond to changing conditions in the domain defined by the objects. An intelligent end-user interface is also provided in the form of read-out tables, graphs, meters, dials and controls.

Bioprocesses employing different operating philosophies will require different specific knowledge (rule-sets) for optimal supervisory control. These "supervisory rule-sets" are used to govern the general operation of the process by identifying growth or production phases and determining which rule-sets are applicable to a given phase (i.e. supervisory rule-sets function as meta rules). The supervisor is responsible for feed scheduling, or on-line modifications to the schedule in response to changes in the process (i.e. faults) and for determining the best time to terminate a reaction and harvest the product. Specific rule-sets are also necessary in order to assess and respond to the metabolic state of a process organism in any given bioprocess (i.e. detect "metabolic faults"). However, for the diagnosis of hardware faults generic rules can be used that apply to all bioprocesses under Bio-SCAN control.

The diagnostic knowledge base has been structured into a number of discrete rule-sets with meta rules again determining which rule-set should be assessed next. The decision to invoke a particular set of rules is based upon a comparison of the current raw, or processed, data with that permitted or expected in the current phase of the bioreaction (characteristic trajectories of any number of variables of interest can be incorporated into the system in addition to constraints). This ensures that knowledge can be focused in response to the occurrence of specific events.

Once a fault has been detected the operator is informed of the problem on a message board and additional rule-sets are invoked in an attempt to establish the nature of the fault. A list of possible causes is then presented to the operator together with advice on how to correct the fault. If Bio-SCAN cannot solve the problem automatically, the operator is requested to confirm a diagnosis (e.g. contamination). Such advice is readily achieved when the process is sufficiently overdetermined, since the knowledge-based system capitalises on redundancies to test for data consistency and error identification. The source of these redundancies is usually balances or load cells, although software sensor estimation of variables reduces the need for hardware redundancy [14, 15].

Since structured, generic rule-sets have been used in the development of the system, coupled with an object-orientated approach to the description of the bioprocess, it is hoped that much of the knowledge base will be generally applicable across a wide range of bioprocesses.

3.3 Knowledge-Based Systems for Bioprocess Scheduling Operations

Commercial scheduling software is, in general, little used in the process industries since it is considered to lack the flexibility and level of detail required for day-to-day plant operation, see DTI Report [16]. This may, in part, be due to the nature of the predominantly algorithmic methods used in these packages (e.g. mixed integer linear programming) which may fail to converge in a reasonable period of time or may require oversimplifications to be made in defining the problem. Moreover, no allowance is made for constraint relaxation, which may be the only method of generating a feasible solution under certain circumstances. The idea of using a KBS approach is attractive in that these systems are well suited to chaining through rules and constraints that govern the heuristic approaches used on most plants. Such an approach may well achieve the speed and flexibility required, albeit at the expense of the near optimal solutions provided by the algorithmic methods [17].

The industrial penicillin installation considered in this paper is a single product plant comprising several production vessels with capacities in excess of 100 000 l, together with numerous storage facilities and a range of smaller seed vessels. Once a seed vessel has been inoculated, the length of the batch can be manipulated, to a certain extent, during the first few hours but then becomes committed to transfer its culture at a fixed time. An appropriate production vessel must therefore be available at this time. Production batches then typically run for around 200 h before harvesting and recovery of product. In order to achieve maximum plant efficiency, production vessels must be utilized in such a way as to ensure high potency media are always available at regular intervals so that downstream processing facilities are continuously supplied. In theory this simply requires an orderly sequence of operations involving seed inoculations, growth of seed, transfer of culture to production vessels and subsequent growth and product formation. In practice, however, the ideal schedule may not be realized because of "upsets" caused by factors such as contaminations of seed or production vessels, maintenance requirements, plant problems, etc. The problem is compounded in that there are different sizes of both seed and production vessels and there may be constraints imposed on where a seed vessel can transfer its culture. This can result in lengthy downtimes for production units if an existing schedule cannot be changed by redirection of previously committed seed cultures.

Further disturbances to the "orderly" harvesting of batches arises as a consequence of the inherent batch to batch variability in production rate. At any one time several batches on the plant may have achieved a sufficient titre to make them candidates for harvesting. However, it is often found that total plant productivity can be improved by harvesting a vessel out of its inoculated sequence. For this reason an on-line forecaster has been developed to identify poor performing batches based on off-line titre information. One method of forecasting is based upon the classification of historical titre profiles and identification of the current batch type to predict future behaviour. An alternative strategy is to employ a curve-fitting approach to a number of possible model structures. Whatever strategy is used, the forecaster seeks to identify which bioreactors should be dispatched to downstream processing as fast as possible and which should be left to run-on. The scheduler must then decide if harvesting a bioreactor out of sequence is workable and economically feasible with regard to the current schedule in operation.

The (simplified) problem description outlined above is not of the conventional flow or job-shop type described in the literature, see e.g. Ku et al. [18] and is essentially one of on-line rescheduling. A solution to the problem using a KBS approach is being developed which combines a model of the plant with a module which will use rule-based heuristics to examine operating strategies. This does not, however preclude the use of algorithmic methods if a mixed approach is found to be needed. The model is being designed to incorporate a detailed knowledge of plant structure and operation and is based on the activity-event representations used in Critical Path Analysis [19]. In this approach however, both activities and events exist as connected objects. Knowledge of how each object exists in the context of plant operation is contained in an associated frame representation. The model will be used to test what-if scenarios by process operators in a manner similar to the system proposed by Bernstein et al. [20].

An "intelligent" user interface will allow operators to easily change constraints or operating conditions and examine the effects. The knowledge base will be particularly useful in analysing the feasibility of the harvesting sequence suggested by the on-line forecaster, and in training operator personnel. System output will be in the form of text messages which can then be manually transferred to the plant planning board.

3.4 Bioprocess Database System

The important problems to be tackled when developing bioprocess databases are not merely how to collect large volumes of different types of data, but also how to validate it before storage and then extract meaningful information. The types of bioprocess data which are usually collected and stored have been reviewed by Carleysmith [21]. In addition, information could also be held concerning characteristic trajectories of key variables with their associated standard deviations as an aid to process evaluation and fault detection by BioSCAN. For the scheduling knowledge base the database would be required to maintain a record of the current plant status with expected completion times of process units (based on the current schedule in operation). Information concerning the maintenance requirements of equipment and vessels could also be incorporated [22], coupled with manpower availability and stock levels of replacement equipment and raw materials. From a user-standpoint the database must also be simple to interrogate, update and view data, and must be flexible in design. Provision for report generation and the statistical analysis of data to establish the relationships between variables would also be useful.

The system chosen to develop the database was INGRES[1], by Relational Technology Inc., which operates a powerful relational database management system [23]. INGRES proved simple to design and update via its form-based-interfaces and supports a powerful query language (SQL) based on virtually any combination of criteria. SQL can also be embedded in external host code, such as "C", which proved important in establishing the link with the scheduling and supervisory knowledge bases.

3.5 System Configuration

The experimental environment which has been developed is shown in Fig. 1. The system is composed of standard industrial process control and signal processing devices, an intermediate bioprocess control computer and a supervisory control computer. The process control and signal processing devices are supplied by Turnbull Control Systems (TCS), the intermediate bioprocess control computer

[1] INGRES is a trademark of Relational Technology Inc.

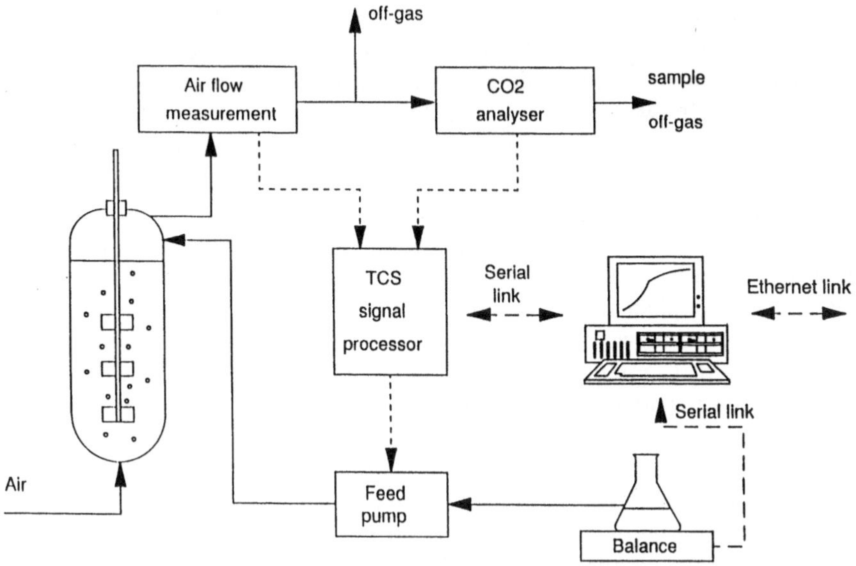

Fig. 1. Experimental environment for bioreactor supervisory control

is an IBM PS2/70 and the supervisory computer is a SUN 3/60 Workstation. The TCS devices provide the interface to the bioprocess under study. They are linked via a multi-drop RS422/RS232 data highway to the bioprocess control computer. This control computer communicates to the supervisory computer via an ethernet network. The whole structure is flexible, reliable and reflects those used in industry.

The bioprocess instrumentation is monitored at a fast sampling rate by the TCS devices, with low level control actions being calculated and implemented directly by the local process controllers and signal processors. The intermediate bioprocess control computer interrogates the TCS devices at regular intervals. The intermediate control computer also carries out signal filtering, data validation and inferential estimation using "soft-sensing" techniques. The conditioned and validated data are then passed to the supervisory computer where it is stored as a file.

On the supervisory computer the data from the intermediate control computer is read by the Bio-SCAN system. This is achieved using the G2 Standard Interface (GSI[2]) which provides facilities for the building of interfaces between G2 and external processes and systems. The data read, and operated upon, by Bio-SCAN is then written back into a file using the same GSI extension process. This file is then read at regular intervals by the intermediate control computer and the information in it is then passed on to the TCS devices.

[2] GSI is a trademark of GENSYM.

For communication between Bio-SCAN and the INGRES database system a different GSI extension process has been developed. Because of the very time consuming nature of relational database management systems it was decided to use a second SUN workstation as the database host machine. This posed an additional communication problem which has been solved by splitting the extension process in two separate processes, namely a Gateway and a Server. The Gateway acts as a true GSI extension process on the Bio-SCAN host machine[3], while the Server runs on the database machine. The processes communicate with each other using the Remote Procedure Call mechanism (RPC[4]) which implements a machine architecture independent way of communication between two computer systems connected by ethernet. Whenever Bio-SCAN needs to interact with the database a request, together with associated information and timing data, is passed via GSI to the Gateway. The Gateway passes this request on to the Server and monitors the progress of the Server by means of the timing information supplied by Bio-SCAN. If for some reason the Server fails to respond within the specified time, the Gateway returns an error message back to Bio-SCAN. Under normal conditions the Server will be able to respond to the request within the specified time. In this case the Gateway returns a completion code and any additional information that may have been sent by the Server. In order to improve robustness the Gateway and Server have been designed to implement a stateless protocol. This means that each request from Bio-SCAN to the database contains all the information that is needed. Successful completion of a request does not depend on any previous or subsequent requests. This allows Bio-SCAN to recover from machine and network failures gracefully.

Both GSI interfaces described above have been implemented in "C" using the hooks provided by GSI which are in the form of "C" function stubs. The body of these function stubs needs to be written by the GSI user. Whenever Bio-SCAN/G2 needs information from the GSI interfaces, it will instruct the extension process to execute the appropriate "C" function. The file interface and the Gateway are both examples of this approach. The Server has also been written in "C". However, rather than being controlled by the GSI, it is controlled by a RPC and the stateless protocol mentioned previously. For the database operations, an embedded SQL has been used.

3.6 System Operation

The supervisory knowledge-based system (KBS), Bio-SCAN, has been designed to oversee the operation of a range of (potentially different) bioprocesses

[3] It is not strictly necessary to run an extension process on the same machine as the G2 real-time expert system. G2 and GSI use the Intelligent Communications Protocol (ICP) which allows extension processes to be run on different host computers and to communicate via ethernet.

[4] RPC is a trademark of Sun Micro Systems.

simultaneously, by providing a permanent, consistent level of "expert" supervision and monitoring. Figure 2 indicates the structure of the software environment.

When a seed or production reaction is to be initiated, requests will initially be made to the database for retrieval of operating conditions and feeding regimes for the strain of organism and mode of operation specified by the scheduler. Once a reactor is running, data from the process is received by Bio-SCAN at regular intervals and is analysed with respect to both expected and recent values (there is provision for limited short-term storage of data within the KBS), and for deviations from "optimal" behaviour. In checking for data consistency, the supervisor also validates the data before it is passed to the database for permanent storage using the protocol detailed above. During the course of a run, requests may also be made for retrieval of, for example, a standard profile of alkali addition or CO_2 evolution to evaluate bioreactor performance. When faults are detected and diagnosed, text messages will be sent to the database specifying the nature of the fault and the action taken. Should a contamination occur, a message is also sent to the scheduler which will then pass back an instruction to stop feeding and terminate the run when an available slot in downstream processing becomes available. Once a batch has been completed, data collected during the course of the run is integrated into standard profiles and abstractable data, such as the specific production rate, can be derived.

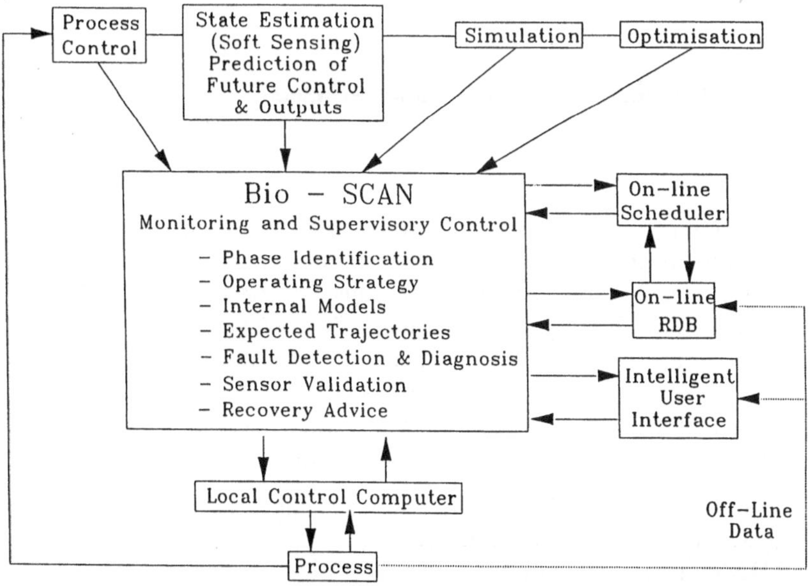

Fig. 2. Supervisory monitoring and control software environment

3.7 Integration with Process Analysis and Control Software

Whilst the knowledge base described above is expected to be capable of adequately supervising bioprocesses under most circumstances, the utility of the system can be greatly extended by interfacing to conventional real-time algorithmic methods of analysing and controlling bioreactions. For this reason a toolkit of on-line modules is being developed that will incorporate the following features into the overall supervisory control strategy:

- Measured signal conditioning, statistical screening and interrogation to place confidence bounds around data.
- Simulation using process models in imported subroutines.
- Identification of process model parameters using on-line recursive parameter estimation algorithms.
- Estimation of hidden states using 'soft sensors'.
- Prediction of future process outputs given pre-specified open-loop feeding regimes. Use of adaptive-predictive control schemes to achieve closed-loop profiling of bioprocess states.
- On-line constrained optimisation using on-line identified process models that capture the essential non-linear structure of the bioprocess.
- Continuous statistical monitoring and process control to provide interactive assessment of overall bioreaction performance against pre-specified specifications.
- Neural Network modelling of non-linear bioprocess models for hidden state estimation, fault detection, and predictive control.

4 Application of Real-Time Knowledge-Based Bioprocess Supervision

Results are presented using data monitored from an industrial fed-batch penicillin process. In these bioreactions the respiratory quotient is usually high and steady and so CO_2 evolution rate or O_2 consumption rate may be used for data analysis. Figure 3 compares the typical CO_2 profiles observed in two "normal" industrial fed-batch penicillin reactions with those of from "abnormal" runs. For proprietary reasons the abscissa scaling in this figure has been omitted. An example of a very simple rule which will detect the possibility of an abnormality in the bioreaction is of the general form:

> IF the CO_2 production rate or rate of change of CO_2 production is greater or less than that expected
> THEN there is evidence of a fault.

Clearly this rule needs to be expanded to incorporate the limits at which the rule

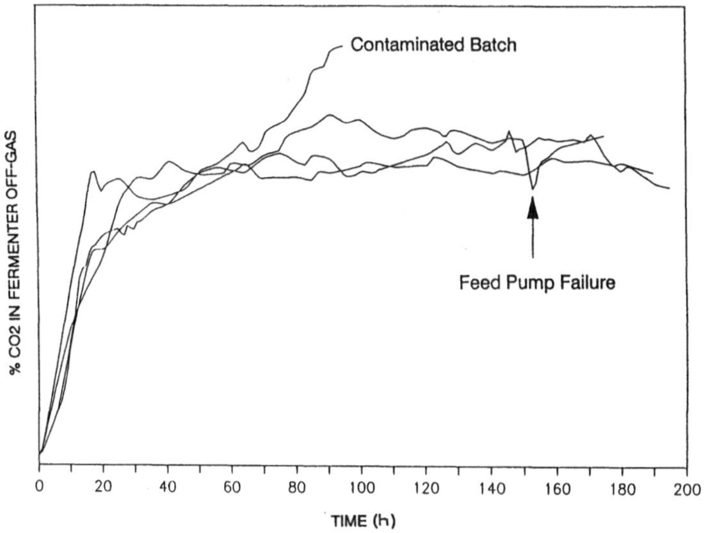

Fig. 3. Carbon dioxide evolution rate test profiles (process Faults)

will fire, how these limits could change during the course of a reaction and perhaps the probability of a process fault. The CO_2 must also be reconciled with other process measurements such as the oxygen uptake rate and dissolved oxygen concentration before diagnostic rule-sets are subsequently invoked. In the contaminated bioreaction shown in Fig. 3, Bio-SCAN detected the increased CO_2 production rate early in the batch. Since this reading was found to be reconcilable with other measurements and no other abnormalities in the operation of the process could be detected, the system concluded that a contamination had most likely occurred and informed the operator after 17 h. It is interesting to note that this contamination was actually detected by the actual process operators after 95 h when laboratory analysis identified the presence of a yeast and terminated the reaction. Had the reaction been terminated earlier a considerable saving in substrate feed costs and plant operation costs would have been realized.

In the case of the second abnormal reaction shown in Fig. 3, Bio-SCAN detected an atypical fall in the CO_2 measurement after 150 h which was not reconcilable with any mini-harvests. As all other operational parameters appeared normal, the system invoked a rule-set which established from balance measurements that substrate flow into the vessel was below normal. The operator was advised to investigate the cause of this problem (e.g. a blocked line or pump failure). Once the problem has been corrected Bio-SCAN is "informed" via the intelligent user interface. Supervisory rules then modify the feeding strategy in an attempt to bring microbial growth back onto a pre-defined trajectory and recover the productivity of the reaction.

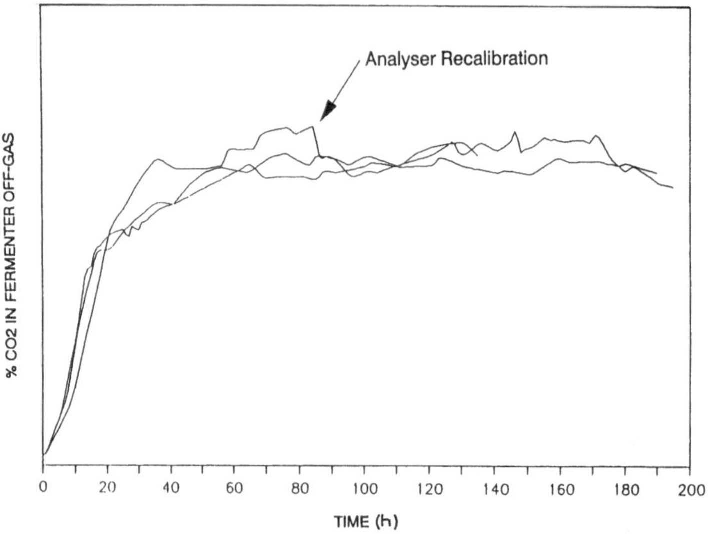

Fig. 4. Carbon dioxide evolution rate test profile (analyser Recalibration)

Figure 4 compares the original two normal bioreactions with a batch during which the CO_2 analyser was detected to have drifted off calibration. Once the suspected calibration error had been confirmed it was corrected at time 81 h and normal bioprocess operation resumed.

5 Bioprocess State Estimation

In contrast to academic research environments, most industrial bioprocess control policies are based upon the use of off-line assay information for process operator supervision. This usually involves the removal of samples from the bioreactor, laboratory analysis and finally operator (or control system) action to correct any undesirable process condition. This philosophy has several important consequences for the control of the system. Off-line sample analysis, due to the staff and laboratory support, is usually a costly operation. A common outcome being that the sampling frequency is reduced so as to minimize costs. This can result in poor process regulation with an inability to react quickly to any process disturbances. The problem of slow sampling is further compounded by the delay induced in the measurement due to analysis. Process operators acting on old information can further degrade the quality of bioprocess supervision. A major industrial requirement would therefore appear to be an algorithm which reduces the off-line analysis frequency whilst at least maintaining, preferably improving, the quality of the information available to the bioprocess operator. The philosophy being that more frequent and more up-to-date information enables improved process operation.

Numerous techniques for the estimation of important "hidden" bioprocess variables papers have appeared in the literature since the early 1980s. The methodologies developed have, however, yet to be proved completely reliable and find acceptance with the industrial community. Industrial systems are inherently non-linear and it is expected that the use of a process model which reflects this essential non-linear structure would prove to be most beneficial. As a consequence the estimation algorithms developed should either be inherently based upon the non-linear structure or alternatively approximate the non-linearities using a linear model with adaptation. Standard non-linear estimation techniques are available, such as the extended Kalman filter [24]. They can, however, suffer from numerical problems and convergence difficulties. Despite these complications many successful applications have been reported, see e.g. Svrcek et al. [25]; Stephanopoulos and San [26, 27].

Several examples were presented at the 1st IFAC Workshop on Modelling and Control of Biotechnical Processes [28] in 1982, e.g. Dekkers [29]; Swiniarski et al. [30], and the following IFAC Symposium held in the Netherlands in 1985 [31], e.g. Ghoul et al. [32]; Montague et al. [33]. A clear message from these early studies was that although model based estimation methods could achieve quite reasonable performance, varying growth regimes and ill-known process disturbance and noise characteristics could invalidate the process description. Much of this early work concentrated on an extended Kalman filter approach coupled with a mechanistic model, the accuracy of which played a major determining factor in the quality of estimation. Deficiencies in the process model, such as unmodelled process dynamics and model parameter variations, have to a certain extent been overcome by applying adaptive estimation techniques, e.g. Leigh and Ng [34]; Shioya et al. [35]; Chattaway and Stephanopoulos [36]. In some cases estimation of the model parameters and process state variables simultaneously is often necessary in order to obtain reliable results.

A prerequisite for any model based estimation technique is the development of a balanced mathematical model of the system under study. The model must be balanced in the sense that a compromise must be made between the conflicting requirements of dynamic complexity and the ability to obtain the parameters of the model either through estimation or experimental measurement. Several techniques have been suggested by Dochain [37] for the development of bioprocess observers, as well as controllers [38]. The observers developed are either of fixed parameter form, partially adaptive form or fully adaptive. The experimental validation of an observer for the on-line state estimation of bioreactor performance (biomass and product) has recently been presented by Dochain et al. [39].

Whilst a large number of process models are available, two major problems are faced. Firstly, the parameters of the models are usually difficult to determine experimentally. Secondly, and perhaps more importantly, the simplifications made in any process model usually result in a need to adapt the model parameters to compensate for dynamic model inaccuracies (model-process mismatch). One approach that has been applied to the penicillin process in order to provide an estimate of biomass concentration [40], is to estimate the parameters of a CER model off-line and use the resulting model in conjunction with an on-line estimator. This results in a partially adaptive observer. That is, a priori knowledge of some of the model parameter values is required for observer application, although it is possible to estimate some of the parameters on-line. In such an observer, the error in the prediction of CER acts as the driving force for state and parameter updating. Although convergence and stability proofs have been developed for observers of the form described, providing certain key criteria are met, partially adaptive observers tend to be insensitive to initial conditions [37]. Adaptive observers that do not require any prior knowledge of the model parameters are known as being fully adaptive. Here the parameters are assumed to be unknown and are identified along with the state estimation.

An alternative starting point for the development of adaptive estimation algorithms is to adopt the philosophy that the bioprocess dynamics can be represented by a general linear input-output model. This approach is similar to that adopted for the development of well known adaptive control laws, except that an additional term representing the secondary process measurement is now included. In the bioprocess plant the secondary measured variable, v, is CO_2 in bioreactor off-gas, the infrequently measured controlled variable, y, is biomass concentration assay, and the manipulated variable variables, u, are the bioreactor feeds. The algorithm thus provides a means by which frequent estimates

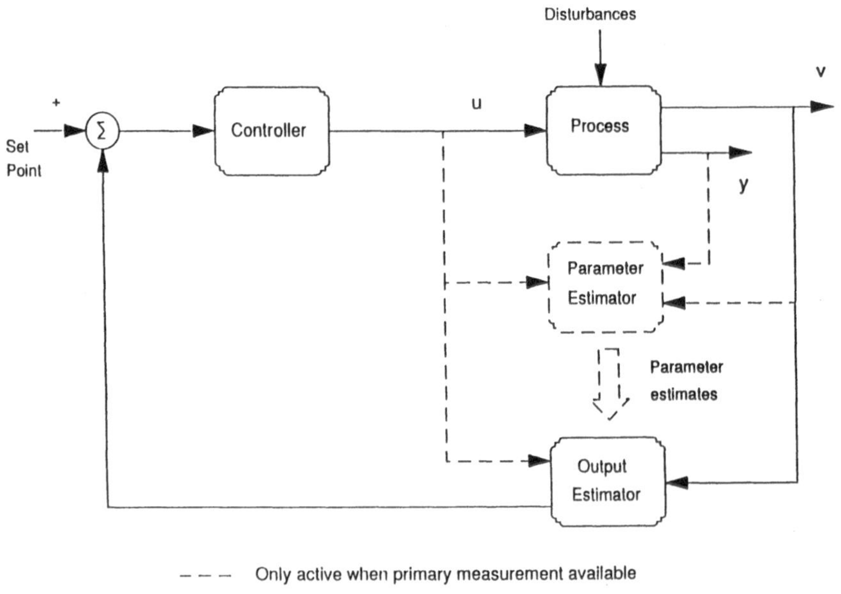

Fig. 5. Generalised linear model based estimator

of biomass can be obtained (at the rate of off-gas measurement) without laboratory assay delay hence improving the information available to the process operator. A representation of the estimation scheme is shown in Fig. 5. The estimator has been applied to the industrial continuous bioprocess so that given infrequent, and irregular, laboratory dry weight data, and frequent CO_2 off-gas analysis and dilution rate data, it is able to predict hourly biomass concentrations. Further details of the derivation of the adaptive inferential state estimator, together with some industrial applications, are set out in Guilandoust et al. [14] and Tham et al. [15].

6 Artificial Neural Network Based Estimation

An alternative learning technique now beginning to be applied is that of neural networks. These are dynamic systems composed of highly interconnected layers of "simple" neurone like processing elements. The potential of modelling process engineering systems with neural networks was initially discussed by McAvoy et al. [41] and Bhat et al. [42]. Results from their studies indicated a tremendous potential of neural networks for process modelling.

Whilst many artificial neural network architectures have been proposed one structure has been predominant; that is the feedforward network. A feedforward neural network is made up of interconnected nonlinear processing elements, termed nodes. Scaled data is presented to the network at nodes of the input layer from where it is propagated to the output through intermediate layers. Each connection has associated with it a scalar weight which acts to modify the signal strength. The nodes in the hidden and output layers perform two functions. In the most simple implementation, the weighted inputs and a bias term are summed and the resulting summation is passed through a sigmoidal activation function.

Following specification of the network inputs, outputs and topology, the network is trained by successive presentations of input-output data pairs. During the training, the data is propagated forward through the network, which adjusts its internal weights to minimize the error between the true output and the output produced by the network. The minimisation of the error can be achieved by the application of suitable optimisation algorithms. After training the network weights are fixed and the model verified on test data prior to actual application. Further details concerning the development of artificial neural network based models can be found in Willis et al. [43].

It is the topology of the network, together with the activation function which provides the potential for non-linear system approximation. Although the network possesses a very rudimentary structure, studies have claimed that any continuous non-linear function could be approximated by a network, with a topology consisting of two hidden layers [44, 45]. These proofs are of major significance, in that an artificial neural network with the appropriate topology can theoretically be used to model any system. To date a few successful general process and bioprocess applications of this modelling philosophy have been reported, e.g. Bhat et al. [46], Lant et al. [47], Thibault et al. [48], Willis et al. [43].

7 Applications of Neural Networks to Bioprocess Supervision

As discussed earlier, a possible solution to the control problems encountered with the mycelial process is to provide the process operators with a more frequent measure of biomass concentration, say once an hour. On-line gas analysis of carbon dioxide concentration provides the opportunity to obtain these process variable estimates at an acceptable frequency. Using the Generalised linear estimator it has been shown that the gas analysis can be used in combination with off-line biomass assays to estimate biomass concentration [47]. The fundamental assumption of the estimator is that the relationship between the process variables can be modelled by a linear time series. Although in practice the process is non-linear, it is assumed that the dynamics can be tracked by on-line adaptation of the linear process model. However, if an improved process model was available then enhanced performance might be expected.

With this aim in mind, a feedforward neural network was considered as a possible vehicle for process model development. Carbon dioxide evolution rate (CER) and reactor dilution rate have already been identified as important variables in biomass estimation. Whilst other variables affect the biomass/CER/dilution rate relationship, for example pH, temperature etc., tight environmental regulation maintains a low variance in these variables. Network inputs are therefore selected to be CER and dilution rate, with biomass concentration being the network output. Not only is the current value of CER fed to the network but also hourly and two hourly delayed values. This time history serves to remove the necessity of dead-time specification, often a problem in model construction. Six network inputs were specified with one process output – the biomass concentration. The network topology consisted of one hidden layer containing three neurones. The operation of the feedforward network estimator can be related to that of the linear estimator, discussed in the previous section, but with fixed parameters (i.e. non-adaptive). Referring to Fig. 5, the neural network learning phase can be related to the "off-line" identification of model (or estimator) parameters. Network prediction of biomass concentration, given measurements of bioprocess feed rate and CER, is similar to biomass estimation of a fixed parameter output estimator. With the neural network, however, the "model" used for estimation is fixed parameter and non-linear, rather than being based upon an adaptive linear model. The quality of the model fit produced by the neural network approach can be seen from the plots in Fig. 6, where the step-like curve is the laboratory (off-line) biomass assay. The bias that can be observed between the biomass assay results and the neural network estimate, from around 25 h onwards, is due to a drift in the on-line CO_2 analyser calibration. This was corrected at around 170 h. At around 210 h a major plant steriliser failure occurs which takes about 5 h to correct. It is encouraging to observe that the neural network estimator remains reliable over the period that this large process disturbance is effective. The variable frequency of the laboratory assays is also quite evident.

The continuous bioprocess considered is required to operate as closely as possible to pre-defined "steady state" operating conditions. In this respect a linear model approximation was acceptable in the vicinity of normal process operating conditions. A fed-batch bioprocess, however, presents more of a problem in terms of model development; the system passes through a wide range of dynamic situations never achieving a steady state.

The measurement problems experienced in controlling the penicillin process are similar to those experienced in the continuous bioprocess. It is desirable to control the biomass growth rate at a low level to optimise the production of penicillin. Whilst the growth rate can be adjusted by varying the rate of substrate (feed) addition, it is unfortunately not possible to measure the rate of growth on-line. The aim therefore is again to develop a model which relates an on-line bioprocess measurement to the process variable of interest. In this case CER was used as the major variable from which to infer biomass concentration. It is worthy of note, however, that in other studies the oxygen uptake rate (OUR) has been used equally successfully. (In the penicillin process after the first 20 h or so, the ratio of CER to OUR, known as the respiratory quotient (RQ), remains constant at around 0.9.)

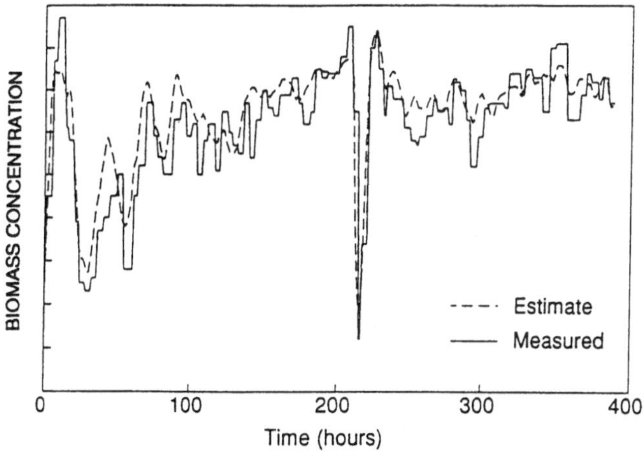

Fig. 6. Neural network non-linear bioprocess model

Process analysis reveals that a combination of three network inputs are required – Carbon dioxide evolution rate (CER) measured in the air flow leaving the reactor, substrate feed, and bioreaction age (i.e. time). Since there are in this case two primary substrate feeds, utilized in a varying ratio, then it is useful to separate the feed in its two components. The network topology was therefore specified as being made up of four inputs, two hidden layers and one output. The determination of the number of nodes in the hidden layers was based upon the optimal method proposed by Wang et al. [45]. In this case four nodes per layer was found to be appropriate. The network was then trained on two sets of process data (not shown).

The quality of the neural network model estimator performance was assessed by testing the network on four sets of "unseen" data. This is demonstrated by Figs. 7, 8, 9 and 10. Here, the estimates of biomass have been obtained solely with the network inputs. Neural network estimates are compared against interpolated off-line biomass measurements. Although the biomass assay results are corrupted with a high level of noise, it can be seen in Figs. 7 and 8 that the neural network estimate is representative of the underlying process behaviour under two quite different sets of operating conditions. This indicates that the neural network model has been able to capture the non-linear process characteristics spanned by the measured data. Figure 9 demonstrates the performance when a contamination occurs towards the end of the batch, at around 120 h, causing the process characteristics to change. The reaction was allowed to run on in order to assess the performance of the neural network estimator over the complete batch. Not surprisingly the network model is no longer applicable to the system.

Figure 10 shows the estimator producing acceptable information until around 60 h into the batch. At this point the estimates are observed to diverge

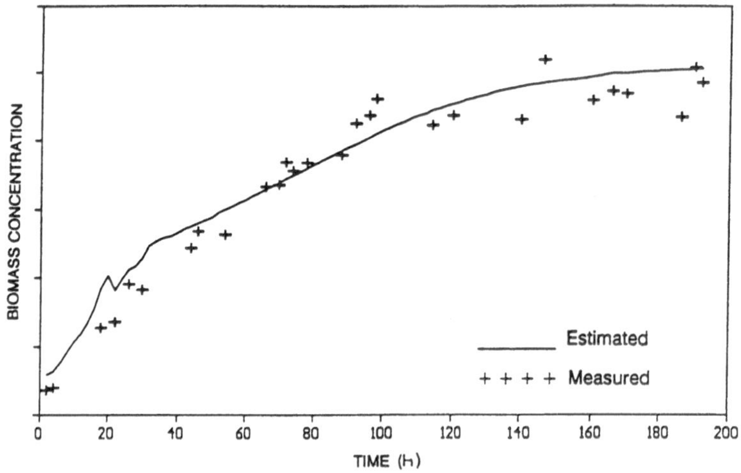

Fig. 7. Neural network penicillin biomass estimation – Test Data Set 1

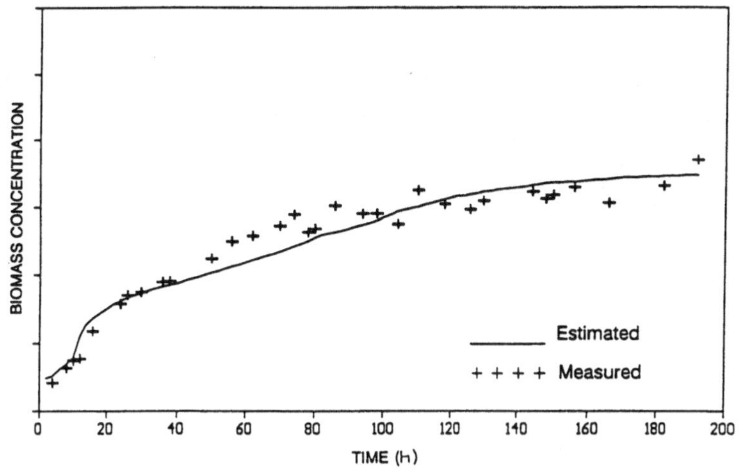

Fig. 8. Neural network penicillin biomass estimation – Test Data Set 2

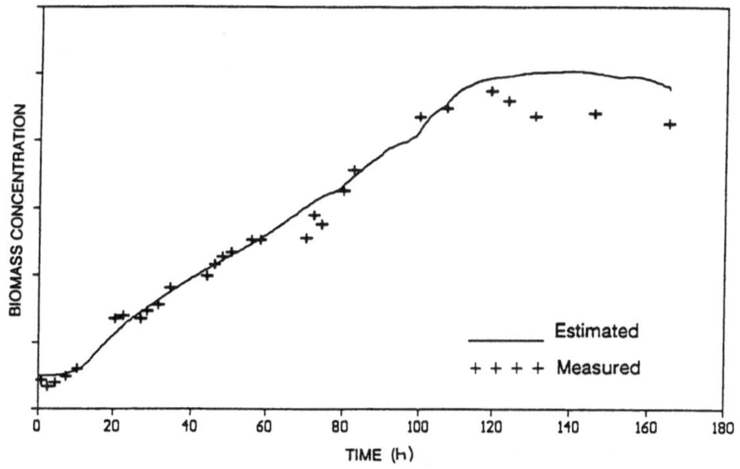

Fig. 9. Neural network penicillin biomass estimation – Test Data Set 3

from the laboratory assays. The operators, having established confidence in the network estimator, were led to check the calibration of the on-line CO_2 analyser. The analyser calibration had drifted resulting in a significant error in the predicted biomass. This was corrected at around the 100 h point. Towards the end of the reaction (140 h) the estimated biomass concentration again diverges. In this case off-line laboratory assays confirmed a contamination. This type of behaviour, considered alongside other recent work in process engineering, suggests that with appropriate reconciliation of data there is potential for

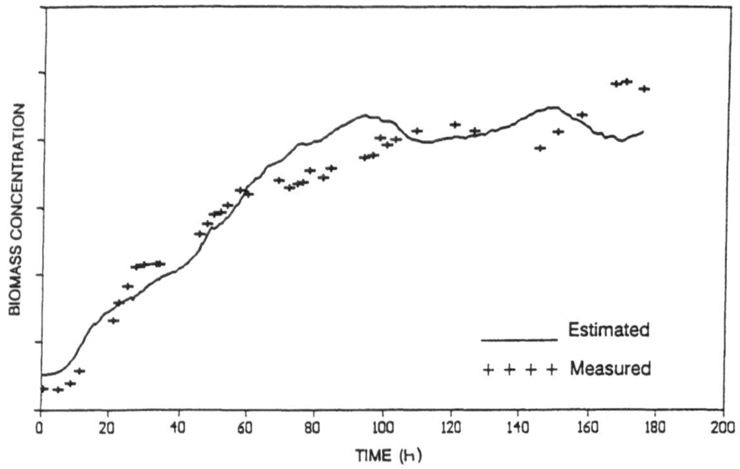

Fig. 10. Neural network penicillin biomass estimation – Test Data Set 4

use in fault detection. Bioprocess faults can of course be associated with reactor and sensor hardware as well as the bioreaction itself.

Acceptable biomass concentration estimates have been achieved without the need for corrective action from off-line assays. It is possible, however, to make such updates. In a similar vein, it is also possible to make use of a first-principles model of the penicillin process in concert with a neural network. In such a scenario the network is used to take account of the process/first-principles-model mismatch. Studies are on-going in these areas.

8 Discussion

The initial objective of the RTKBS study outlined in this paper has been the development of a generally applicable real-time knowledge-based system for the supervisory control of bioprocesses. The supervisory strategy also performs sensor validation, fault detection and diagnostic tasks and provides advice to aid productivity recovery following fault correction. In order to achieve this aim, generic-type rules have been used as much as possible, particularly in the case of hardware fault detection and diagnosis. Process specific knowledge can be incorporated within the existing framework relatively easily owing to the modularity and object orientated approach adopted in the development of this system.

Further work is underway aimed at incorporating on-line modules to perform process simulation, soft-sensing and control. Intelligent use of this information should significantly improve the overall performance of the system in providing for the continuous on-line optimisation of bioprocesses. Most recently, studies into the use of knowledge-based methodologies in bioprocess scheduling have been started.

Model-based bioprocess state estimators have tended to use relatively simple process models. It is of course possible to develop quite complex first principles bioprocess models. Such an approach is capable of delivering a considerable improvement in the bioreaction data available for supervision and control well beyond that provided by existing techniques. However, these benefits must be weighed against the considerable effort which must be employed to model biochemical systems and the resulting high cost of doing so. Care needs to be excercised, however, in that inappropriate assumptions inevitably lead to problems in observer application.

A new and exciting approach, that of neural network modelling, provides a generic non-linear modelling technique. Network models have been developed for the on-line prediction of intersample values of low frequency biomass measurements using higher frequency secondary measurements. The potential of these networks to model dynamic non-linear bioprocesses, in order to provide an on-line state estimator, has been demonstrated by application to two industrial bioprocesses. Although there are a number of open issues such as optimum network size, etc., the power of this new technique should be judged by its ability to achieve acceptable state estimation and process control in a cost effective manner.

Acknowledgements: The authors would like to acknowledge the support of Gensym Corporation; Smith Kline Beecham, ICI Biological Products and Marlow Foods, in terms of access to plant and process data and financial support; and also the U.K. Science and Engineering Research Council for their financial support.

9 References

1. Stephanopoulos G, Tsiveriotis C (1988) Towards a systematic method for the generalisation of fermentation data. Preprints, 4th Int Cong on computer applications in fermentation technology, Cambridge, UK, September
2. Karim MN, Halme A (1988) Reconciliation of measurement data in fermentation using on-line expert system. Preprints of 4th International Cong on computer applications in fermentation technology, Cambridge, UK, September
3. Cooney CL, O'Conner GM, Sanchez-Rivera F (1988) An expert system for intelligent supervisory control of fermentation processes. Preprints of 8th International Biotech Symp, Paris, July
4. Hofmeister M, Halasz L, Rippin DWT (1989) Comp Chem Eng 13: 1255
5. Halme A (1987) Measurement and estimation in bioreactors. Proc International Symposium on control of biotechnical processes, University of Newcastle, UK, October
6. Montague GA, Morris AJ, Ward AC (1989) Fermentation monitoring and control: A perspective. Biotechnology and genetic engineering reviews, vol 7, December
7. Edelman J, Fewell A, Solomons GL (1983) Myco-protein – a new food. Nutrition abstracts and reviews in clinical nutrition, Series A, vol 53, p 6471

8. Chantler MJ (1988) Real-time aspects of expert systems in process control. In: Expert systems in process control, IEE Colloqium, March 1988
9. Stephanopoulos G, Stephanopoulos G (1986) Artificial intelligence in the development and design of biochemical processes. Trends in Biotechnology, p 241
10. Moore RL, Hawkinson LB, Levin M, Hofman AG, Matthews BL, David MH (1987) Expert systems methodology for real-time process control. Proceedings of the 10th IFAC World Congress, Munich, FRG, July
11. Qi C, Wang S-Q, Wang J-C (1988) Application of expert system to the operation and control of industrial antibiotic fermentation process. Preprints of 4th International Congress on computer applications in fermentation technology, Cambridge, UK, September
12. Moore R (1988) The G2 real-time expert system for process control. International symposium on advanced process supervision and real-time knowledge-based control, University of Newcastle, UK, November
13. Aynsley M, Peel D, Morris AJ (1989) A real-time knowledge-based system for fermentation control. Proc ACC, Pittsburgh, USA, June
14. Guilandoust MT, Morris AJ, Tham MT (1987) Adaptive inferential control. Proc IEE, vol 134, Pt D
15. Tham MT, Montague GA, Morris AJ, Lant PA (1991) Soft-sensors for process estimation and inferential control. J Proc Cont, vol 13–14
16. D.T.I Report (1989) Batch process scheduling computer software. The R&D Clearing House
17. Bruno G, Elia A, Laface P (1986) IEEE Computer 32: 32
18. Ku H-M, Rajagopalan D, Karimi I (1987) Scheduling in batch processes. Chem Eng Prog, August, p 35
19. Barnetson P (1969) Critical path planning. Butterworth, London
20. Bernstein G, Carlson EC, Felder RM (1989) Development of a simulation-based decision support system for a multipurpose pharmaceutical plant. AIChE Annual Meeting, San Francisco, CA, November
21. Carleysmith SW (1988) Data handling for fermentation development – An industrial approach. Preprints, 4th Int Cong on computer applications in fermentation technology, Cambridge, UK, September
22. Flynn DS (1982) Instrumentation for fermentation processes. Proc 1st IFAC Conf, modelling and control of biotechnical processes, Helsinki, Finland
23. Codd EF (1970) Comm ACM 13: 377
24. Anderson BDO, Moore JL (1979) Optimal filtering. Prentice Hall, New Jersey
25. Svrcek WY, Elliot RF, Zajic JE (1974) Biotech Bioeng 16: 827
26. Stephanopoulos G, San KY (1981) State estimation for computer control of biochemical reactors. Advances in biotechnology, vol 1, p 399
27. Stephanopoulos G, San KY (1984) Biotech Bioeng 26: 1176
28. Halme A (ed) (1982) Proceedings 1st IFAC workshop on modelling and control of biotechnical processes, Helsinki, August, Pergamon, Oxford
29. Dekkers RM (1982) State estimation of a fed-batch bakers yeast fermentation. Proc 1st IFAC workshop, modelling and control of biotechnical processes. Helsinki, Finland, August
30. Swiniarski R, Lesniewski A, Dewshi MAM, Ng MH, Leigh JR (1982) Progress towards estimation of biomass in a batch fermentation process. Proc 1st IFAC workshop, modelling and control of biotechnical processes, Helsinki, Finland
31. Johnson A (ed) (1985) Proceedings 1st IFAC symposium on modelling and control of biotechnical processes, Noordwijkerhout, December, Pergamon, Oxford
32. Ghoul M, Pons MN, Engasser JM, Border J (1985) Extended Kalman filtering technique for the control of *Candida utilis* production. 1st IFAC symposium modelling and control of biotechnical processes, Noordwijkerhout, p 165
33. Montague GA, Morris AJ, Wright AR, Aynsley M, Ward A (1985) Parameter adaptive control of the fed-batch penicillin fermentation. 1st IFAC symposium modelling and control of biotechnological processes, Noordwijkerhout, p 39
34. Leigh JR, Ng MH (1984) Estimation of biomass and secondary product in batch fermentation. 6th International Conference on analysis and optimisation of systems, Nice, France, p 19
35. Shioya S, Shimizu H, Ogata M, Takamatsu T (1985) Measurement of state variables and controlling biochemical reaction processes. 1st IFAC symp modelling and control of biotechnical processes, Noordwijkerhout, p 49
36. Chattaway T, Stephanopoulos G (1987) A new technique for estimation in bioreactors: adaptive

state observers. Proc international workshop on control of biotechnical processes. University of Newcastle, UK, October
37. Dochain D (1986) PhD thesis. University of Louvain, Belgium
38. Bastin G, Dochain D (1988) Non-linear adaptive control algorithms for fermentation processes. Proceedings of American Control Conference, Atlanta, USA, p 1124
39. Dochain D, de Buyl E, Bastin G (1988) Experimental validation of a methodology for on-line estimation in bioreactors. 4th ICCAFT Conference – modelling and control of biotechnical processes. Cambridge, UK, September
40. Di Massimo C, Saunders ACG, Morris AJ, Montague GA (1989) Non-linear estimation and control of mycelial fermentations. Proceedings of the American Control Conference, Pittsburgh, USA, p 1994
41. McAvoy T, Wang N, Bhat N (1989) Use of neural networks for interpreting biosensor data. Proceedings of the American Control Conference, Pittsburgh, USA
42. Bhat N, Minderman P, McAvoy T, Wang N (1989) Modelling chemical process systems via neural computation. International symposium – control for profit, University of Newcastle, UK, November
43. Willis MJ, Di Massimo C, Montague GA, Tham MT, Morris AJ (1991) Artificial neural networks in process engineering. Proc IEE, Pt D, 138, No 3, p 256–266
44. Cybenko G (1989) Approximations by superpositions of a signoidal function. Math. Cont. Signal and System, 2, p. 303–314
45. Wang Z, Tham MT, Morris AJ (1992) Multilayer feedforward neural networks: Canonical form approximation of nonlinearity, International Journal of Control
46. Bhat N, Minderman P, McAvoy TJ (1989) Use of neural nets for modelling of chemical process systems, Preprints IFAC Symp Dycord + 89, Maastricht, The Netherlands, Aug. 21–23, p 147
47. Lant P, Willis MJ, Montague GA, Tham MT, Morris AJ (1990) A comparison of adaptive estimation with neural network based techniques for bioprocess application. Proceedings of the American Control Conference, San Diego, USA, p 2173
48. Thibault J, Van Breusegem V, Cheruy A (1991) Biotech Bioeng 36: 1041

Host-Vector Interactions in *Escherichia coli*

James E. Bailey
Department of Chemical Engineering, California Institute of Technology,
Pasadena, CA 91125, USA

Dedicated to Prof. Karl Schügerl on the occasion of his 65 birthday

Introduction of a DNA vector into *E. coli* for the purposes of cloned gene expression can perturb native cell functions at many levels. The presence of foreign DNA can alter regulation of chromosomal DNA replication, regulation of transcription of chromosomal genes, ribosome functions and RNA turnover, activities of regulatory proteins, chaperone and protease levels, membrane energetics and protein post-translational processing, as well as energy and intermediary metabolism of the cell. The combined effects of these interactions on the metabolic characteristics of the host-vector system have major implications for yields of cloned biotechnological products and interactions of genetically engineered organisms with their environment.

Advances in Biochemical Engineering
Biotechnology, Vol. 48
Managing Editor: A. Fiechter

Symbols and Abbreviations

C	time for chromosomal replication
D	time between replication form termination and cell division
$G_{\beta L}$	number of β-lactamase gene copies per cell
IPTG	isopropylthiogalatoside
$k_{dm\beta L}$	specific degradation rate of β-lactamase message
$[mRNA]_{\beta L}$	intracellular concentration of β-lactamase messenger RNA
μ	specific growth rate
η_{tx}	transcription efficiency of the β-lactamase gene
t	time

1 Introduction

The genome of an organism contains the genes and regulatory elements which interact and function to establish the chemical, physical, and functional characteristics of that organism. "Wild-type" or "normal" attributes of the organism are a consequence of these interactions and functions. In seeking to clone a segment of DNA or to express a cloned gene, introduction of recombinant vectors into the bacterium *Escherichia coli* is a common procedure. The insertion of new DNA into the cell can perturb and in some instances severely disrupt the normal balance of activities found in a wild-type cell. This review considers experiments and mathematical modeling which have been conducted to elucidate the types of perturbations which can arise.

Studies of the interlocking influences of vector and host cell upon each other have major implications in biotechnology and basic science. For example, when manufacturing a cloned therapeutic protein such as granulocyte colony stimulating factor using genetically engineered *E. coli*, the reduction in specific growth rate caused by high-level expression of the product must be taken into account when designing the host-vector expression system and when operating the process. From a basic science perspective, careful study of the metabolic and regulatory characteristics of cells containing a recombinant DNA vector offers insights into native cell function in the sense of a stimulus-response experiment. Alterations in host cell metabolism which result from introduction of a cloning vector are also of basic importance in the study of metabolic engineering. When seeking to discern the influence on cell metabolism of expressing an active cloned molecule in the cell, it is essential to recognize and to control for other influences on cell functions which result from all other interactions with the cloning vector.

This review will emphasize general points and will not endeavor to catalog every experiment or study involving a different vector or host strain. Because all data on interactions between hosts and vectors and their consequences are obtained from in vivo studies, it is difficult if not impossible to dissect clearly the most critical loci of the interactions. Many different types of perturbations and changes in cell functions occur simultaneously. Nevertheless, in an effort to emphasize the diverse ways by which introduction of new DNA can and does alter cell function, the following comments will begin with the discussion of perturbation of functions of host cell DNA, progress through perturbations in RNA and protein functions and processing, and subsequently turn to other special areas of metabolism before arriving at the most common subject of vector effects on specific growth rates.

The key concept is the parasitic and perturbing influence of the added vector. However, it is critically important that this parasitism be recognized as a complex network of interconnections with many aspects of cell function and not something as simple as "a demand for extra DNA synthesis" or "a demand for more ATP". Considering alterations in the bacterial genome in a somewhat more general context, it is clear that there is a complete spectrum of responses possible ranging from a silent mutation which changes the wild-type DNA without influencing cell function whatsoever to the introduction of a lytic virus which completely preempts native cellular activity resulting in virus synthesis and cell death. Thus, a broad spectrum of responses must be considered, and the cellular responses will vary based upon the particular characteristics of the vector and the host involved. Except where otherwise noted, this review will emphasize plasmid vectors, but many of the same comments and considerations apply as well to phage vectors.

2 Interactions with DNA Functions

The most direct possible interaction of a cloning vector with the DNA of the host and its functions occurs when introduction of the vector modifies the host genome. This occurs for integrating vectors or if the autonomously replicating vector contains a transposon or a region of homology with the host genome such that some insertion is made into the host DNA. Such insertions can have many effects, of course, ranging from silence to lethality by disrupting a function essential for viability. Many other types of influences are possible by inactivating important genes or regulatory elements. Such interactions are commonplace with expression vectors for mammalian cells such that the resulting clones differ widely in their growth characteristics, cell size, productivity, and other phenotypic characteristics, presumably due to different integration sites and their subsequent consequences for host cell function. Such difficulties can be avoided in *E. coli*, for example, by using vector and selection technology which localizes the integration site such that a function unimportant for the growth conditions of interest is targeted. Such a manipulation, for example, would involve integrating a gene into the *lacZ* locus for a strain which would be grown on a glucose-containing medium. Other types of interactions which can also exert profound effects on the characteristics of the host-vector system are indirect in the sense that the presence of the vector alters the native distribution of DNA-binding proteins on the host DNA.

2.1 Replication

Bacterial plasmid vectors contain origins of replication consisting of regulatory sequences which serve to control initiation of replication of the plasmid and also the site of replication initiation. However, the vector typically does not encode the necessary DNA polymerase and associated enzymes for DNA replication, depending upon the host enzymes for these functions. Further, the plasmid may depend upon other host proteins involved in regulation of chromosomal DNA replication with resultant perturbations on the replication of the host chromosome. No experiments have been designed to test the nature and extent of these types of interactions exclusively. However, available data show significant effects of plasmid presence on DNA and cell cycle regulation in *E. coli*, effects which are likely due to the types of interactions just outlined.

In particular, Seo and Bailey determined the nuclear cell cycle parameters C and D for a plasmid-free host strain and for a set of copy mutant vectors which are extremely similar except for small domains which influence their copy number (number of plasmids per cell) [1]. (Fig. 1 shows the structure of these plasmids which have been employed in many investigations of host-vector interactions which will be discussed in this review.) C is the time required after initiation of chromosomal DNA replication for the corresponding replication fork to reach the terminus of the chromosome. The time interval designated D is the time which elapses from the time the replication fork reaches the terminus until cell division occurs. By analyzing the distribution of single-cell DNA contents in different recombinant strains of *E. coli* using flow cytometry and analyzing those distributions with population model relationships, Seo and Bailey showed that plasmid-carrying strains exhibited significantly smaller C and D times than the corresponding plasmid-free host, *E. coli* HB101 (Table 1). These results are interesting and somewhat unexpected in the sense that, because of the general parasitic and disruptive effect of plasmid presence, one would tend

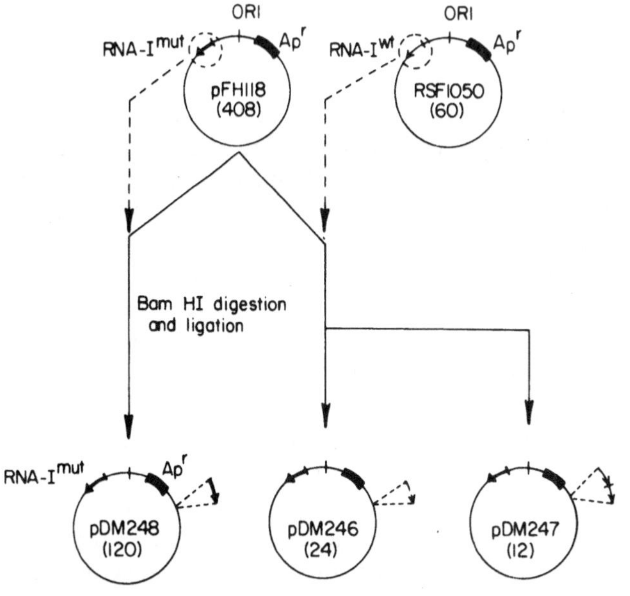

Fig. 1. Geneology of the plasmids used in many studies of host-plasmid interactions. Indicated in parentheses is the copy number (plasmids/cell) of each plasmid in *E. coli* HB101 grown in Luria-Bertani (LB) medium. Alternative nomenclature for these plasmids is based on these reference copy numbers (e.g., P24 denotes pDM246, P60 denotes RSF 1050, etc.). Reprinted from [24]

Table 1. Estimated C and D periods for recombinant *E. coli* HB101 strains. Estimates were obtained by two independent methods. Flow cytometry analysis of the single-cell DNA content distribution following chloramphenicol (CM) addition combined with theory enabled estimation of the number of active replication forks in an exponentially growing population and thereby, based on population theory, the C and D times. Direct comparison results were obtained by least squares optimization of C and D in the theoretical steady-state single-cell DNA distribution compared to corresponding flow cytometry data. Growth conditions: LB medium at 37 °C at pH 7.0. Reprinted from [1]

	CM-treatment			Direct comparison		
Strain	C (min)	D (min)	C + D (min)	C (min)	D (min)	C + D (min)
HB101 only	36.8	19.9	56.7	37.3	20.5	57.8
HB101:pDM247	30.3	15.4	45.7	31.0	15.5	46.5
HB101:pDM246	30.9	16.7	47.6	31.8	15.0	46.8
HB101:RSF1050	33.3	16.2	49.5	33.8	13.7	47.6
HB101:pDM248	32.1	15.0	47.1	33.5	14.6	48.1

to expect increases in these cell cycle times rather than decreases. Furthermore, it is extremely interesting that the plasmid copy number exerts almost no effect on the magnitude of the change of C and D; merely the presence of plasmids, in this case all plasmids with a ColE1-type origin of replication, is sufficient to give rise to these responses in the host-vector construct.

2.2 *Transcription*

E. coli RNA polymerase apoenzyme and holoenzyme bind nonspecifically to DNA. Accordingly, introduction of additional DNA into the cell as a multicopy plasmid can be expected to cause significant parasitic partitioning of RNA polymerase onto the plasmid DNA. Of course by simple mass action relationships, this parasitic partitioning would not be significant for a small, low copy plasmid but could be substantial for a large or high copy vector. A model of cloned gene expression and its regulation which considers the influences of nonspecific binding on vector DNA on availability of polymerase for gene transcription clearly indicates substantial effects when plasmid DNA content of the cell approaches the DNA content of the host cell genome (corresponding to around 1000 copies of a vector of 4 kb [2]).

Moreover, plasmids carry several promoters. For example, in pBR322 there are two promoters associated with the origin of replication (for the small transcripts RNAI and RNAII) as well as the promoters associated with the ampicillin resistance and tetracycline resistance genes. In a construct designed for high level expression of a cloned protein, promoters of such great activity are employed that they may, even when present at relatively low levels, compete strongly for available RNA polymerase holoenzyme.

These disturbances in the distribution of RNA polymerase relative to that in the wild-type cell are expected to influence the level of expression of various host

Table 2. Determinations of relative levels of total RNA and ribosome components in *E. coli* HB101 carrying different plasmids from Fig. 1 (P0 denotes the plasmid-free strain). Reprinted from [3]

Strain	RNA[a]/protein ($\mu g\ \mu g^{-1}$)	70S[b] (AUFS per 10^3 dpm)	50S[c] ($70S^{-1}$)	30S[c] ($70S^{-1}$)
P0	0.26	–	–	–
P24	0.25	3.6	0.23	0.16
P60	0.28	3.7	0.19	0.14
P120	0.23	2.3	0.14	0.09

[a] Calculated with Ab_{260} of 1.0 to be equivalent to 40 μg RNA per ml.
[b] Normalized with dpm of ^{3}H-labelled cells and Ab_{420} of P24 AUFS denotes absorption units full scale.
[c] Height ratio from absorbance (254 nm) scan of ribosome fractions separated in a sucrose gradient

cell genes. Experiments to examine ribosomal RNA contents and relative levels of ribosome components in cells carrying different members of the family of plasmids shown in Fig. 1 have clearly shown the consequences of such perturbations [3]. As plasmid copy number increases, RNA and active 70S ribosome levels first decline slowly, then drop rapidly (Table 2). These trends parallel those for specific growth rate discussed below.

Wood and Peretti also observed a decline in ribosomal RNA (rRNA) levels for increasing copy number in experiments employing the plasmids in Fig. 1 in *E. coli* HB101 [4]. Interestingly, their detailed study showed that this decline in steady-state rRNA levels with increasing plasmid content was a consequence of an increase in rRNA degradation which more than compensated for an increase in rRNA synthesis rate with increasing copy number. A different trend was observed in studies of the consequences of induction of the strong *tac* promoter in a pUC-based expression plasmid [5]. In this case rRNA content increased as the cloned promoter was induced because the increase in rRNA synthesis rate with increasing induction dominated a smaller increase in rRNA degradation rate.

If the pattern of individual proteins found in the cell is regarded as indicative, at least in part, of the corresponding pattern of transcription, the alterations in cellular content of many individual host cell proteins which arise upon introduction of multicopy plasmids indicate significant perturbations in host cell gene expression patterns which presumably result from shifts in distribution of RNA polymerase. Based on computer image analysis of two-dimensional gel electrophoresis patterns of recombinant *E. coli* HB101 extracts, levels of many host cell proteins are altered by presence of multicopy plasmids (Table 3) [3]. Levels of stress response proteins detected are generally higher, while many metabolic enzymes exhibit reduced levels at the highest copy number studied. Lower activities of glucose-6-phosphate dehydrogenase, fructose 1,6-diphosphate aldolase, and fructose 1,6-diphosphotase were observed in *E. coli* DH5α carrying the higher copy members of the plasmid family in Fig. 1 [6].

The appearance of foreign or misfolded native proteins in the cell has been suggested to be the primary signal for the stress or "heat shock" response of the

Table 3. Relative amounts of individual proteins in plasmid-free (P0) *E. coli* HB101 and in HB101 carrying P60 and P120 as determined by analytical two-dimensional protein gel electrophoresis. MW is molecular weight (kDa) and pI is isoelectric pH. Results for P0 indicate the fraction of that protein relative to all proteins detected. P0 numbers denote ^{35}S spot intensity on the basis of a total intensity of 10^6 for all spots detected on the P0 gel. The values for P60 and P120 strains were normalized by the corresponding values for P0 to indicate changes in protein levels relative to the plasmid-free host. A value of 1.0 thus indicates no change of protein level in comparison to the plasmid-free strain P0. Reprinted from [3]

Name	M_r (kDa)	pI	Strain		
			P0[a]	P60[b]	P120[b]
Carbon metabolism					
Oxoglutarate dehydrogenase complex					
Oxoglutarate dehydrogenase	110.0	6.36	1314	2.68	0.92
Dihydrolipoate succinyl-transferase	48.0	5.74	2239	2.55	1.14
Dihydrolipoate dehydrogenase	49.3	6.25	2716	1.72	0.81
PEP Carboxylase	92.9	5.88	2151	0.99	1.64
Pyruvate dehydrogenase complex					
Pyruvate dehydrogenase	99.6	5.70	4652	1.13	0.60
Dihydrolipoate transacetylase	74.7	5.12	10,593	0.90	0.75
Dihydrolipoate dehydrogenase	49.3	6.25	2716	1.72	0.81
Pyruvate kinase I	53.8	6.23	8631	0.97	0.65
Succinate dehydrogenase	71.3	6.17	199	3.73	2.20
Succinate thiokinase (β)	41.8	5.66	1270	2.26	1.09
Amino acid metabolism					
Aspartate aminotransferase	41.7	5.93	945	0.75	1.35
Glutamine synthetase,					
Adenylated	50.7	5.52	2721	0.93	1.51
Unadenylated	50.8	5.57	1167	1.23	0.34
Nucleotide metabolism					
Aspartate transcarbamoylase	33.9	7.00	1556	2.03	0.57
Carbamoyl-phosphate synthetase (β)	136.2	5.57	1754	1.20	1.77
Protein synthesis translation					
Aspartyl-tRNA synthetase	66.1	5.75	1278	0.96	0.25
Elongation factor					
G	87.5	5.60	11,475	1.35	0.67
Tu	43.2	5.71	103,503	0.71	0.63
Ribosomal subunit proteins					
S1	69.1	5.00	8608	–	0.58
S5	17.2	–	21,326	1.15	0.45
S6a	15.0	5.70	2195	2.02	0.08
S6b	15.1	5.63	10,567	1.07	0.31
S6c	15.2	5.53	7674	0.60	–
L1	34.4	–	6358	1.14	0.45
L3	28.7	–	10,163	1.12	0.26
L7/12	12.2	4.91	28,168	0.92	0.84
L11	14.7	–	26,411	0.85	0.28
L25	11.0	–	12,972	0.81	0.69
Heat shock proteins					
Dna K	71.3	4.96	7506	1.14	1.12
Htp,					
G	69.3	5.24	563	1.52	2.05

Table 3. (continued)

Name	M_r (kDa)	pI	Strain P0[a]	P60[b]	P120[b]
H	35.9	5.65	355	0.59	1.33
Grp E	25.1	4.96	1889	1.02	2.27
Gro EL	58.0	5.01	13,770	1.16	1.63
Other					
β-Lactamase[c]	29.6	5.77	–	2.19	8.52

[a] This is ppm as dpm (specific protein) per dpm total valid protein spots.
[b] This is ppm per ppm P0.
[c] This is normalized with respect to P24 which has 2889 ppm

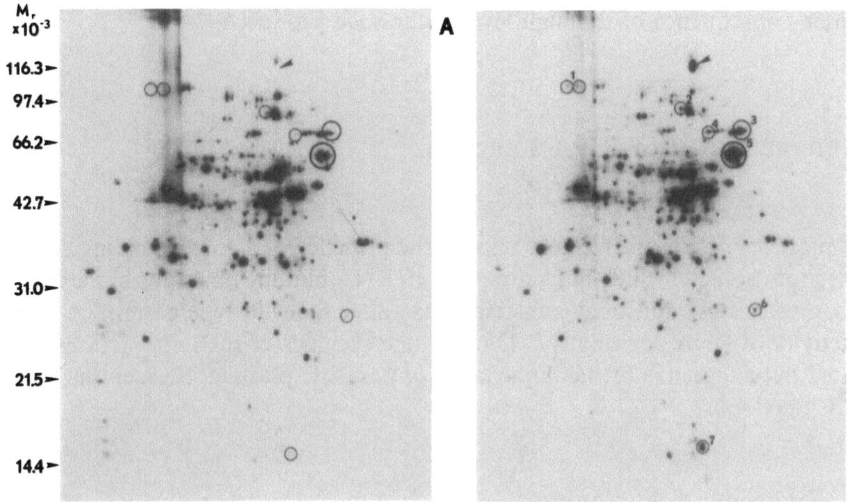

Fig. 2A,B. Two dimensional gel electrophoresis of *E. coli* BGF1 before (**A**) and after (**B**) the induction of the CSH11 mutant β-galactosidase. Cells were grown at 37 °C and pulse-labeled before (A) and 30 min after (B) IPTG addition. The following stress proteins were identified according to known databases (Phillips et al., 1987): (*1*) Protease La (H94.0, H94.1); (*2*) HtpM (F84.1); (*3*) DnaK (B66.0); (*4*) HtpG (C62.5); (*5*) GroEL (B56.5); (*6*) GrpE (B25.3); and (*7*) GroES (C15.4). The position of the abnormal β-galactosidase is indicated with an *arrow*. The right side of the gels corresponds to lower pH values. Samples loaded each had 100000 cpm total. Reprinted from [9]

cell [7]. Thus, the appearance of a heterologous or mutant homologous protein resulting from cloning and expression of its gene is expected to activate stress response gene expression. Indeed, this has been observed in experiments with several different recombinant strains of *E. coli*, including the experiments summarized in Table 3. Another study has revealed the sensitivity of the stress response to small quantities of mutant protein. M. Kosinski constructed *E. coli* BGF1 by inserting into the chromosome of a wild-type strain (MG1655) a single copy of a *lacZ* gene containing the CSH11 mutation [8]. Addition of IPTG to a culture of BGF1 resulted in elevated expression of several stress response proteins as determined by analytical two-dimensional gel electrophoresis (Fig. 2) [9]. As a warning on the difficulties of designing experiments which clearly delineate only host-plasmid interactions, the recent discovery that addition of IPTG to wild type *E. coli* alters metabolism and protein pattern should be noted [10].

One possible strategy for reducing alterations in host cell gene transcription resulting from parasitic partitioning of host cell RNA polymerase is introduction of a separate polymerase-promoter pair for cloned gene expression. Indeed, this strategy is embodied to some extent in the expression strategy which employs phage T7 polymerase and the corresponding phage T7 promoter for cloned gene expression [11]. However, this promoter is so strong that cell specific growth rate is reduced following induction, presumably as a direct and indirect consequence of the high level of message produced.

3 Interactions with RNA Functions

The presence of plasmid vectors implies the presence of a corresponding set of transcripts, some of which may interact with RNA binding functions of the host. Also, presence of additional transcripts may alter the wild-type pattern or level of activity of RNA degradation. Data on the specifics of such interactions are limited but suggestive of the importance of parasitic plasmid interactions with RNA functions.

3.1 Ribosome Partitioning

In the detailed single-cell model of plasmid-carrying *E. coli* presented by Peretti and Bailey [12], competition for active ribosomes between host cell messages and plasmid-derived messages was considered as was the impact of such competition on the total level of ribosome activity in the cell. Simulations, which will be discussed in greater detail later, indicated potential significance of this interaction, motivating an extensive subsequent study of a special recombinant expression configuration employing specialized ribosomes.

Introduced by DeBoer and coworkers [13], introduction of a cloned gene for a mutant 16S ribosomal RNA and corresponding modification of the Shine-Dalgarno(SD) sequence on a messenger RNA enables construction of a subpopulation of specialized ribosomes which are competent only to translate the correspondingly modified messages. Although the synthesis and assembly of these specialized ribosomes is certainly a nontrivial perturbation of host cell metabolism, the establishment of a separate class of ribosomes for translating plasmid message in principle could reduce the deleterious effects of high levels of plasmid message and of strong ribosome binding sites on those messages on translation of host-cell message. An extensive investigation of this strategy and its consequences has recently been presented by Wood and Peretti [14]. They observed a 30% reduction in specific growth rate of the culture following expression of the specialized 16S RNA; however, no further decline in growth rate was observed following induction of transcription of cloned β-galactosidase message bearing the modified SD sequence.

3.2 RNA Degradation

Measurements of the rate of β-lactamase message synthesis and degradation have been conducted for *E. coli* HB101 carrying several of the plasmids of Fig. 1. Estimation of the message-specific mRNA degradation rate was based upon the following material balance for β-lactamase mRNA concentration per unit volume of cells averaged over the entire culture:

$$\frac{d[mRNA]_{\beta L}}{dt} = \eta_{tx} G_{\beta L} - [k_{dm\beta L} + \mu] \tag{1}$$

$\eta_{tx} G_{\beta L}$ is the rate of message synthesis and [mRNA] denotes the intracellular concentration of RNA in mols per unit volume of cells. $k_{dm\beta L}$ is the message-specific mRNA degradation rate constant and μ is the specific growth rate of the culture. For either a chemostat culture in steady state or a batch culture in exponential balanced growth, the message concentration is expected to achieve a steady state obtained by equating the left hand side of Eq. (1) to zero and allowing an explicit calculation of the degradation rate constant in terms of other parameters.

After measuring the relative steady state β-lactamase message levels using Northern hybridization and the relative overall rates of β-lactamase message

Table 4. Kinetic data for synthesis and degradation of β-lactamase messenger RNA in *E. coli* HB101 carrying the different plasmids shown in Fig. 1. The first column of data shows relative synthesis rates. Division of these by the corresponding plasmid copy numbers gives the relative transcription efficiencies η_{tx} indicated. The relative β-lactamase message specific degradation rate constants $k_{dm\beta L}$ were estimated from Eq. (1) making use of other measurements of relative steady-state β-lactamase mRNA levels. Reprinted from [15]

Strain	Counts per min[a] (10^{-6})	η_{tx} (10^{-8} cpm per gene per min)	$k_{dm\beta L}$ (h^{-1})
P12	0.17	0.89	21
P60	0.72	0.75	2.8
P120	6.6	3.4	79
P408	690	110	330

[a] Amount of labeled β-lactamase mRNA on a per cell basis. Total RNA was pulse-labeled with [^{3}H] uracil, hybridized to nitrocellulose bound β-lactamase DNA and digested with both RNase A and RNase T_1 prior to scintillation counting

synthesis using pulse-chase methods, estimates for the relative β-lactamase message degradation rate constants for *E. coli* HB101 carrying four of the plasmids in Fig. 1 were obtained (Table 4) [15]. There is a major increase in message degradation rate at the highest copy numbers. Similar trends of greater degradation rates for rRNA with increasing copy number [14] and with increasing cloned promoter induction [5] have been observed. Although the basis for the significantly increased rate of RNA turnover in these recombinant cultures is unknown, the effect is striking and must be considered as one of the major consequences of plasmid presence.

4 Interactions with Protein Functions

Several influences of plasmid presence on the functions of proteins involved in regulating and conducting DNA, RNA, and protein synthesis have already been indicated. Further interactions of plasmids and their products with host-cell protein functions are discussed next.

4.1 Regulatory Proteins

As noted earlier, plasmid replication often depends upon particular proteins which are also implicated in controlling host-cell chromosome replication initiation. In addition, expression plasmids are generally designed to afford controlled and regulated expression of the product gene. Often, the proteins required for regulation of these cloned promoter-operator systems are provided by the host. Introduction of many copies of the cloned promoter-operator may exhaust the available supply of regulatory protein, resulting in loss of control of expression of the cloned product gene and also the corresponding gene (or genes) in the host genome. Such loss of control is well known and well documented for multicopy plasmids carrying the *lac* and *tac* promoter-operators. However, similar effects can also be expected for other promoters which depend upon host-cell proteins for their regulation.

4.2 Chaperones and Proteins of the Secretory Apparatus

Host cell chaperones are implicated in proper folding and assembly of numerous host cell proteins, and also in preserving export-competent states of host-cell polypeptides prior to membrane translocation [16, 17]. Expression of cloned proteins may result in their interaction with host-cell chaperones, thereby modifying the distribution of chaperones on host proteins and altering their folding, processing, and export. As noted earlier, genetic introduction of either multicopy plasmids or a small quantity of mutant protein into the cell activates expression of stress response proteins including several chaperones. Besides perturbing host cell protein metabolism, these interactions may also influence the folding and targeting of the cloned protein.

4.3 Protein Turnover

Expression of a cloned protein may influence the level and distribution of protease activities in the cell. Clear demonstration of this is provided in recent experiments by Kosinski and coworkers which show major responses of cellular proteolytic activity to expression of a very small quantity of a mutant *E. coli* β-galactosidase from a single chromosomally integrated gene [9]. In particular, the activity of ATP-dependent proteases was found to increase upon induction of mutant β-galactosidase expression while the level of ATP-independent protease activity was relatively unaffected. This finding is significant in view of the implication of the ATP-dependent protease La in the initial, rate-limiting step of intracellular proteolysis [18].

5 Interactions with Membrane Functions

5.1 Processing of Secreted Proteins

Cloned proteins may be targeted for export through the cytoplasmic membrane by gene fusion of a signal sequence with the product gene. Although this strategy has not proven generally effective (only those proteins secreted in their native hosts appear to be reasonable candidates for secretion in a heterologous host), many such constructs have been employed effectively to obtain export of a cloned protein into the periplasm of *E. coli*. In some cases, use of such fusion constructs with strong promoters has resulted in lethality following induction. This so called "overproduction lethality" has been attributed to interference of the cloned preprotein with processing of host preproteins. That is, by mass action, the large quantity of cloned preprotein occupies much of the limited capacity of the host cell secretory apparatus, thereby impeding normal rates of host cell preprotein processing and transport. Thus, there is again a parasitic interaction and competition between the newly expressed cloned preprotein and the normal processing of host cell secreted proteins. Another consequence of this type of interaction may be reduction in integrity of the outer membrane of *E. coli* since its proteins are not processed sufficiently rapidly (relative to the cell division rate) at the cytoplasmic membrane. This can result in leakage of periplasmic proteins into the medium [see, for example, 19]. The latter phenomenon can be exploited to advantage in practice to produce cloned products extracellularly using an *E. coli* expression system. Whether overproduction lethality and preprotein processing limitations might be circumvented by overexpressing cloned proteins of the host secretory apparatus provides intriguing challenges for future research.

5.2 *Influences on Membrane Energetics*

Axe and Bailey have reported an in vivo nuclear magnetic resonance phosphorous-31 spectroscopy study of host and plasmid-carrying *E. coli* [20]. Observations of intracellular and extracellular pH transients following addition of glucose suggest a difference in membrane energetics between the transformed and unmodified cells. Whether these changes result from differences in proton pumping activities or differences in susceptibility to passive leakage of proteins is not known. Also, it should be noted that the plasmid involved in these experiments encoded pre-β-lactamase, a protein new to the cell which must be processed at the cytoplasmic membrane. Whether the observed effects on transmembrane pH difference is connected with the expression of this cloned preprotein is not established but worth considering.

6 Interactions with Energy and Precursor Metabolism

Alterations in the metabolic characteristics of host-cells transformed with plasmids relative to the unmodified hosts are often ascribed somewhat casually to additional demands for ATP or additional demands on precursors for DNA synthesis. The alterations which plasmid presence causes in ATP and DNA synthesis for use in biomass production can be well estimated by rigorous methods. Stoichiometric analysis of bacterial growth pioneered by Stouthamer and coworkers [21, 22] can be extended to analysis of host-plasmid systems. Because of limited data available on the detailed composition of plasmid-bearing strains, the stoichiometric analysis conducted by Da Silva and Bailey assumed the host-cell biomass composition remained unaltered by plasmid presence. Plasmids, plasmid-derived transcripts, and proteins encoded on the plasmid were assumed additive to the host-cell composition. Then, following the approach of Stouthamer, calculations were undertaken for a variety of scenarios to estimate the effect of the presence of plasmids and the expression of plasmid genes at different levels on the amount of ATP required for growth [23]. Some of the results of these calculations are summarized in Table 5.

As might have been expected given the ATP demands for protein synthesis in the wild-type host [21], it is the extent of plasmid-directed protein synthesis which determines whether plasmid presence significantly increases the ATP demand on the cell. From this stoichiometric viewpoint, even propagation of

Table 5. Theoretical estimates of the maximum cell yield from ATP ($Y^{max, ATP}$) and of ATP requirements for synthesis of different classes of cellular components. Six cases are considered for recombinant cells. The parameters n and δ denote the number of plasmids per cell and the fraction of total protein synthesized which is encoded on the plasmid, respectively. The "product retained" figures include plasmid-encoded protein as part of total cellular protein; the "product secreted" figures do not. Thus the latter figures may also be interpreted as requirements and yields for *active* biomass. Reprinted from [23]

		Recombinant cell					
		Product retained			Product secreted		
ATP requirement for synthesis of	Normal cell	$\delta = 0.2$ $n = 50$	$\delta = 0.5$ $n = 50$	$\delta = 0.5$ $n = 100$	$\delta = 0.2$ $n = 50$	$\delta = 0.5$ $n = 50$	$\delta = 0.5$ $n = 100$
Polysaccharide	20.52	18.12	13.45	13.44	20.49	20.49	20.46
Protein (f)	13.55	11.96	8.88	8.87	13.53	13.53	13.51
(p)	191.42	169.02	125.48	125.36	191.13	191.13	190.85
Lipid	1.40	1.24	0.92	0.92	1.40	1.40	1.40
RNA (f)	34.50	30.46	22.62	22.59	34.45	34.45	34.40
(p)	9.20	8.12	6.03	6.03	9.19	9.19	9.17
DNA (f)	8.64	7.63	5.66	5.66	8.63	8.63	8.61
(p)	1.92	1.70	1.26	1.26	1.92	1.92	1.91
Product (f)	0.00	2.99	8.88	8.87	3.38	13.53	13.51
(p)	0.00	42.26	125.48	125.36	47.78	191.13	190.85
Plasmid DNA (f)	0.00	0.35	0.26	0.53	0.40	0.40	0.80
(p)	0.00	0.08	0.06	0.12	0.09	0.09	0.18
mRNA Turnover	13.80	12.19	9.05	9.04	13.78	13.78	13.76
Sub-total	294.95	306.12	328.03	328.03	346.16	499.66	499.41
Transport of							
NH_4^+	42.42	44.60	48.84	48.85	50.43	74.40	74.37
K^+	1.92	1.69	1.26	1.26	1.92	1.92	1.91
P	7.75	6.88	5.11	5.13	7.78	7.78	7.82
Total ATP requirement	347.04	359.30	383.24	383.27	406.29	583.76	583.51
Y_{ATP}^{max} (g cells per mol ATP)	28.8	27.8	26.1	26.1	24.6	17.1	17.1

very high copy number vectors does not make much difference on global energy metabolism. However, according to this stoichiometric analysis, perturbations in ATP requirements of the cell can become important when the expression level of the cloned proteins becomes significant relative to that for all host cell proteins. The fact that substantial growth rate reductions occur with high copy plasmids and with moderate to low overall expression of cloned proteins [24] is a clear indication that the interactions discussed in previous sections may dominate plasmid effects on cellular ATP utilization.

This type of analysis of course does not include and cannot estimate the effects of plasmid presence on maintenance. As pointed out by Stouthamer and others, maintenance is a kinetic rather than a stoichiometric process and, in *E. coli*, consists primarily of maintaining the energetic charge of the cytoplasmic membrane [22]. Unfortunately, there have been few definitive studies on maintenance requirements for plasmid-bearing cells compared to the corresponding plasmid-free hosts, probably because of the difficulty of long term continuous culture of plasmid-bearing strains.

Some genetic engineering groups have advocated careful consideration of "codon usage" when designing synthetic genes for high level expression in a heterologous host. This means that, among the available codons, one should choose those which are preferred in that particular host. Data on codon usage in the host of interest are typically obtained by amino acid analysis of some of its most highly expressed native proteins [25]. While the importance of codon usage for achieving high level expression is a subject of ongoing debate, suggestion of its importance implies another level of potential host-plasmid interaction. In particular, if a cloned gene involves unusually high usage of a particular codon which is relatively infrequent in the host, availability of the charged tRNA corresponding to that codon may become limiting and may then influence expression of some host-cell proteins. Clearly, one could identify many other particular loci in the overall metabolic system of the cell at which similar imbalances and perturbations could result as a consequence of plasmid-directed activity.

7 Effects on Specific Growth Rate

Several if not all of the potential perturbations in regulation and activity of cellular metabolic processes caused by plasmid presence occur simultaneously. The most commonly studied and widely reported consequence of plasmid presence is alteration in the specific growth rate of the cell relative to that of the plasmid-free host.

Listed in Table 6 are the specific growth rates in two different media, measured from exponential growth in shake flasks, of *E. coli* HB101 bearing the different copy mutant plasmids illustrated in Fig. 1 [24]. There is a monotonic decline in specific growth rate as the plasmid copy number increases. This decline is gradual for low copy numbers and subsequently becomes more steep. These studies and others have observed the expected consequence of introducing a vector which provides no useful products for the host. The resulting extent of specific growth rate reduction follows the expected trends, increasing with higher level expression of a cloned product [26, 27] and with higher copy number vectors. However, it should be noted that an opposite influence has been observed in the case of *E. coli* expressing the hemoglobin cloned from *Vitreoscilla* [28]. When grown under oxygen-limited conditions, a strain engineered to express *Vitreoscilla* hemoglobin can outgrow a similar strain which does not express hemoglobin [29]. Of course this result can be viewed as a particular instance of widely applied selection strategies designed to provide a growth advantage for the plasmid-bearing strain in a particular growth environ-

Table 6. Plasmid content effects on recombinant *E. coli* HB101 maximum specific growth rates.[a] Reprinted from [24]

Plasmid	Reference copy number	Relative specific growth rates		
		M9	LB	M9C
pDM247	12	0.93	0.92	0.97
pDM246	24	0.88	0.91	0.94
RSF1050	60	0.88	0.87	0.88
pDM248	120	0.78	0.82	0.84
pFH118	408	0.68	0.77	0.82

[a] Values shown are the ratio of measured specific growth rate to specific growth rate of the plasmid-free host in the same medium. *E. coli* HB101 specific growth rates: LB medium: 1.38 h^{-1}; M9C medium: 0.79 h^{-1}; M9 medium: 0.41 h^{-1}

ment. However, in this case the "selective environment" is a condition of oxygen limitation which is commonly encountered in bioprocessing.

8 Inclusion Bodies and "Toxic" Products

Many heterologous and some homologous proteins, when overexpressed, aggregate into microscopic particles called inclusion bodies (IBs) or refractile bodies. These are believed to be produced by interactions of partially folded intermediates in the folding pathway which create an off-pathway diversion into these aggregated structures [30]. It is common for aberrant elongated cell morphologies to occur when extensive inclusion bodies synthesis is observed in recombinant *E. coli*. For example, Fig. 3 shows a microphotograph of *E. coli* carrying the plasmid pRED2 which expresses the *Vitreoscilla* hemoglobin peptide driven by the native *Vitreoscilla* hemoglobin promoter [31] (also called ORE and OxyPro™ in the literature). It is unknown how inclusion body formation is coupled mechanistically to the formation of these elongated cells. Presumably the presence of the inclusion body (or the processes of its synthesis) interferes with transverse cell wall synthesis and other reactions associated with the final stages of the cell division cycle. It is possible also that, by physical interference with native intracellular events, inclusion bodies disturb and distort normal cell cycle progression.

Experience in the biotechnology industry has shown that cell growth ceases or is severely inhibited when certain heterologous proteins are expressed at very low levels. Thus, for an effective process for producing such proteins, it is essential that the protein be expressed from a regulated promoter and that regulation in the inactivated state be extremely tight. Novel promoter regulatory configurations combining minimal preinduction activity and high transcription rates post induction have recently been proposed based on molecular-level mathematical models [32]. Why certain heterologous proteins in very

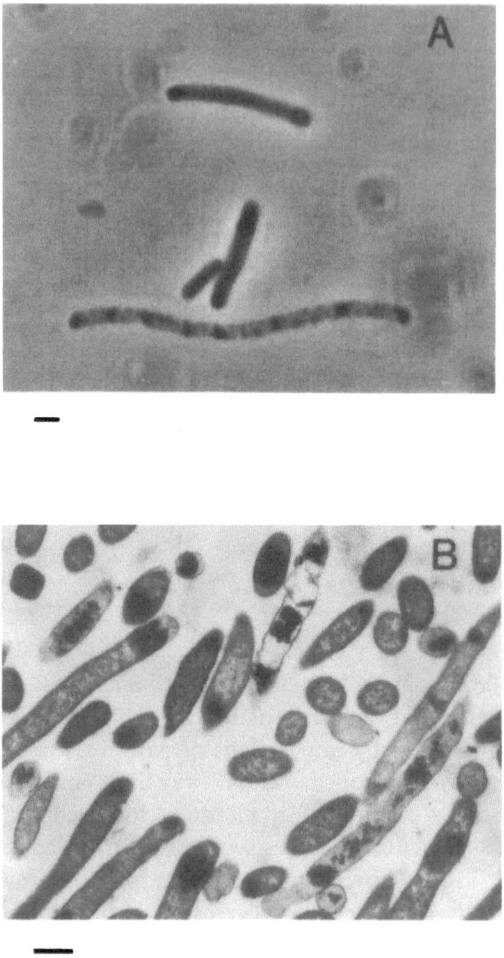

Fig. 3A,B. Photomicrographs of *E. coli* JM101: pRED2 cells. **A**: phase contrast micrograph (bar is 1 μm). **B**: transmission electron micrograph (bar is 1 μm). In addition to refractile-type inclusion bodies, cells also contain another inclusion body morphology designated floccule-type. Reprinted from [31]

small amounts so greatly interfere with cell growth is also not known. Some hypothetical causes include binding of the heterologous protein to crucial DNA or RNA sequences, interaction of the heterologous DNA with chaperones or proteases in the cell, or catalysis by the heterologous protein of reactions which produce toxic components.

9 Implications of Selection Pressure

In order to isolate clones containing vectors of interest or in order to favor growth in culture of the desired recombinant strains, it is common to design a complementary genetic and medium formulation such that cells containing the vector exhibit substantially more rapid specific growth rate than cells lacking the vector. Common selection mechanisms used in this context are antibiotic resistance encoded on a single gene on the vector coupled with addition of this antibiotic to the medium. Another familiar motif is complementation in which the host cell contains a mutation which inactivates a gene encoding an enzyme essential for growth; provision of this gene on the plasmid then complements the host cell defect and allows the vector-containing cells to grow.

It should be recognized, however, that the relief of the lethal situation provided by the marker gene on the vector may not be complete. Thus, in a culture growing in a high concentration of ampicillin in order to select for recombinants containing the β-lactamase gene marker, the presence of the antibiotic may still interfere with cell envelope structure to some degree. Similarly, a cell which is complemented by a "good" copy of the mutated host cell gene may still suffer of some level of "nutrient limitation" or even nutrient excess because the synthesis of that enzyme may not be properly coordinated with the entire cellular metabolic structure as is the wild-type chromosomal copy.

10 Mathematical Models of Host-Plasmid Interactions

Mathematical representation of cellular activities which include some accounting for host-plasmid interactions have been formulated to address different questions and to serve different purposes. As in any modeling of cellular phenomenon, one can never include everything that is happening in the cell which may be important. Instead, it is necessary to choose a particular subset of cellular components and interactions to include in the model. These choices must be based upon the intended use of the model, and, unfortunately, what is included and what is neglected remains a subject of some degree of modeling art. There is no rigorous, theoretical basis for choosing which variables and interactions should be retained in the model in order to describe the essence of the system and to avoid omission of some critically important component or mechanism. The following discussion briefly considers the types of models which have been formulated relating to host-plasmid interactions and applications of such models.

10.1 Small Structured Models

In any description of cell kinetics, simple models which represent the cell by a small number of pseudocomponents are convenient and often, by judicious choice of a small number of parameters, can be adjusted to describe the growth or product formation trajectories of interest. Although the components of such small models are often assigned a loose physical interpretation such as total nucleic acids or total protein or "total synthetic component", these labels are more useful for conceptual guidance in proposing reasonable kinetic forms than they are rigorously valid interpretations. A small structured model is basically an empirical fit to the observed phenomena in which most if not all model parameters have no clear physical relationship with the fundamental parameters characterizing the reactions and equilibria actually occurring within the cell. However, as a trade-off, the numbers of parameters involved and the computational complexity of such models is minimized. Therefore, these types of models are particularly useful for implementation in control schemes and for representing experimental observations of the overall characteristics of the cultures.

The four compartment model of Nielsen et al. includes variables for "active biomass" interpreted to be primarily ribosomes, a lumped component consisting of structural material and chromosomal DNA, plasmid DNA, and cloned protein [33]. This model simulates many of the important qualitative features of host-plasmid interactions including decreasing maximum specific growth rate with increasing plasmid content and variations in RNA concentration and plasmid content with culture specific growth rate. These trends are also well simulated by an eight-compartment model formulated by Bentley and Kompala [34].

10.2 Large-scale Structured Models

Large-scale computer models containing many variables (say of order 30) and parameters (in excess of 100) serve a different purpose. These models represent an attempt to calculate, in a systematic and coordinated fashion, the consequences of many simultaneous interactions within the cell, some of which are represented in great detail, perhaps even at the level of the known molecular mechanism. These models have an important place for hypothesis testing in the following sense: the model in itself is an embodiment of a certain set of variables and corresponding interactions precisely and rigorously defined by the model equations. The model thus defines a self-consistent "computer cell" which possesses its own intrinsic characteristics based entirely on the model's structure and parameter values. Therefore, the model calculations with the computer cell show the extent to which such a structure is capable of manifesting the behavior observed experimentally for biological cells. As such, agreement between model and experiment is an indication that the structure implemented in the model is a plausible one for interpretation of the experimentally observed phenomenon.

However, it is extremely dangerous to extend this logic excessively and to suggest that, because such a detailed model simulates even many aspects of experimentally observed behavior in cells, the model itself is correct in all of its details and that use of such models can be used to identify molecular mechanisms. This is not possible because of possible interactions in the model such that errors in one segment of the model may compensate for other errors in other segments of the model.

A large-scale structured model of plasmid-bearing *E. coli* was constructed by Peretti and Bailey [12, 35] based upon the plasmid-free single-cell *E. coli* model developed by Shuler and several of his collaborators [36]. The Peretti model was built to examine the implications of the types of parasitic interactions shown schematically in Fig. 4. That is, the Peretti model explicitly calculates the quantity of actively transcribing RNA polymerase as well as the quantity of actively translating ribosomes in the cell and includes, in the case of plasmid-bearing cells, the influence of plasmids on these quantities. Figure 5 depicts schematically the components and processes considered in the Peretti–Bailey model for RNA polymerase-promoter interactions and gives some indication of the level of detail involved.

Table 7 lists simulated growth rates and cloned gene product levels of *E. coli* carrying different copy number plasmids and comparable experimental data. Clearly, the trends are captured correctly. This model also simulated reasonably well the unexpected influence of plasmid presence on the C and D cell cycle parameters observed experimentally [1, 12]. Because experiments showing the phenomenon had not yet been accomplished, this model did not include a strong effect of increasing plasmid copy number on decreased cloned gene message stability which was subsequently discovered [15]. Possible major influences of unknown interactions is always a potential pitfall of cell kinetics modeling, but it must be clearly recognized when working with large models which may, by virtue of their complexity, give a falsely assuring impression of completeness and rigor.

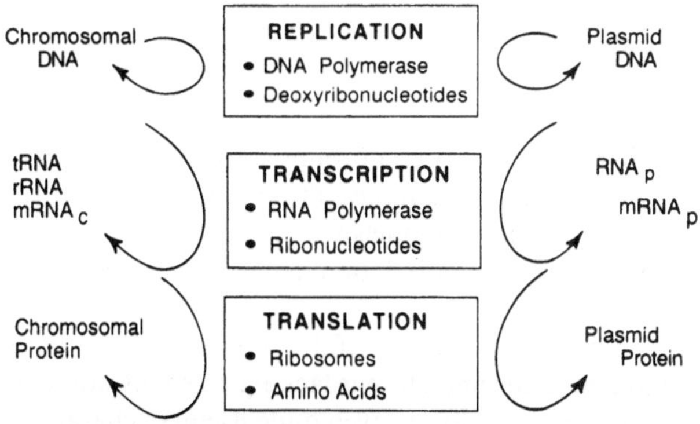

Fig. 4. A schematic representation of the competition between chromosomal- and plasmid-directed macromolecular systems for the cell's macromolecular synthetic machinery. Reprinted from [12]

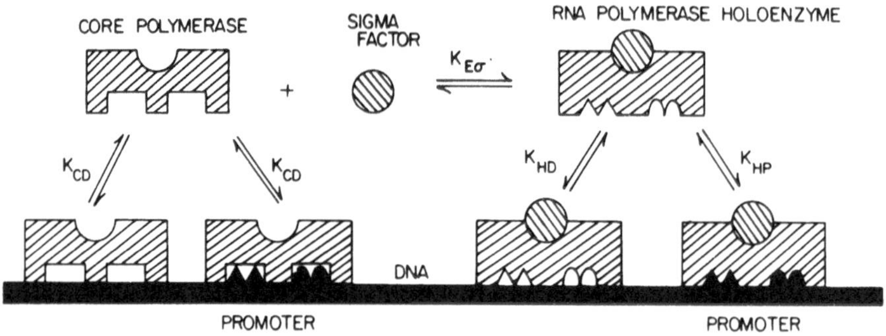

Fig. 5. RNA polymerase exists in many conformations. Binding of the sigma subunit to core polymerase alters binding selectivity of enzyme. Resultant holoenzyme has much greater affinity for promoter sites and decreased affinity for nonpromoter regions relative to core polymerase. These protein and DNA species and their relative affinities are explicitly calculated in the Peretti–Bailey single-cell model for *E. coli*. Reprinted from [35]

Table 7. Comparison of experimental and model simulation results for the specific growth rate and the amount of cloned gene product per plasmid as a function of the relative number of plasmids per cell. Reprinted from [12]

Number of plasmids per cell	Relative specific growth rates		Cloned gene product, plasmid specific synthesis	
	Experimental	Calculated	Experimental[a]	Calculated
0	1	1		
1	0.92	0.94	1	1
2	0.91	0.90	2.1	1.9
5	0.87	0.89	4.2	3.6
10	0.82	0.87	5.4	4.1
20	–	0.66	–	5.2
34	0.77	–	7.0	–

Experimental values are taken from Ref. [24] for *E. coli* HB101 with the plasmid in Fig. 1 grown at 37 °C in LB enriched with leucine, proline and thiamine and supplemented with 0.2% glucose

10.3 Population Models

Productivity of a cloned protein can be greatly compromised by rearrangements of the vector or by "shedding" of the vector from the culture as it grows. Several investigators have formulated mathematical models which seek to describe the changes in the plasmid content of a growing population in batch and continuous bioreactors.

Simplest of these models but yet extremely useful has been an approach which lumps all cells containing plasmids and all cells lacking plasmids into two separate populations. These types models have proven quite useful in characterizing the fundamental parameters of plasmid segregation and the overall qualitative characteristics of the dynamics of plasmid content in bacterial cultures [37, 38]. Other models have appended to this basic structure other possible subpopulations which contain rearranged plasmids [39].

A few population balance models have been formulated and applied to describe plasmid propagation and segregational instability (e.g., [40]). These calculate not only the concentration of plasmid-free cells in the culture but the concentrations in the culture of cells with differing plasmid contents, in fact often over a continuous distribution of plasmid numbers. Such models provide a clear description of the relationship between the intrinsic instability of the plasmid and the corresponding distribution of plasmid contents (including of course the average plasmid content) in the recombinant culture.

11 Concluding Remarks

Although there has been an attempt in the previous presentation to decompose the different elements of host-plasmid interactions, it must be recognized that these several interactions arise simultaneously, greatly complicating any clear interpretations of the mechanisms or dominant phenomena which determine the physiological characteristics of a plasmid-bearing bacterium. The intrinsic complexity of the situation is illustrated by a body of phenomenology which is known to those who have worked with many different combinations of *E. coli* strains and plasmids. For reasons which are not at all apparent, some plasmids are more stable in some strains than others. Also, some proteins are expressed much better in hosts with a particular genetic background than in other hosts. Although of course in some cases hosts have been especially engineered to enhance some important property of DNA stability or protein processing, in many other cases the connection between the details of the host cell genotype and the different physiological responses following introduction of a plasmid are difficult to discern and in some cases almost impossible to conceive. Undoubtedly further research on the physiology of plasmid-bearing cells and on their internal composition and regulatory characteristics will clarify some of these perplexing observations. However, the complexity of the influences of a new piece of DNA on all levels of cellular function is sufficiently great that some trial and error optimization of vector and host will likely remain for some time an integral part of the process to identify an effective expression system.

Acknowledgements: The author's research in the field of host-plasmid interactions has been supported by the National Science Foundation and by the Advancement Industrial Concepts Division of The United States Department of Energy.

References

1. Seo J-H, Bailey JE (1987) Biotechnol Bioeng 30: 297
2. Lee SB, Bailey JE (1984) Biotechnol Bioeng 26: 1383
3. Birnbaum S, Bailey JE (1991) Biotechnol Bioeng 37: 736
4. Wood TK, Peretti SW (1990) Biotechnol Bioeng 36: 865
5. Wood TK , Peretti SW (1991) Biotechnol Bioeng 38: 397
6. Mason CA, Bailey JE (1989) Appl Microbiol Biotechnol 32: 54
7. Ananthan J, Goldberg AL, Voellmy R (1986) Science 232: 522
8. Kosinski MJ, Bailey JE (1991) J Biotechnol 18: 55
9. Kosinski MJ, Rinas U, Bailey JE (1992) Appl Microbiol Biotechnol (in press)
10. Kosinski MJ, Rinas U, Bailey JE (1992) Appl Microbiol Biotechnol (in press)
11. Reznikoff WS, McClure WR (1986) In: Reznikoff W, Gold L (eds) Maximizing gene expression. Butterworths, Boston, Mass, p 1
12. Peretti SW, Bailey JE (1987) Biotechnol Bioeng 29: 316
13. DeBoer HA, Hui A, Comstock LJ, Wong E, Vasser M (1983) DNA 2: 231
14. Wood TK, Peretti SW (1990) Biotechnol Bioeng 36: 865
15. Peretti SW, Bailey JE, Lee JJ (1989) Biotechnol Bioeng 34: 902
16. Rothman JE (1989) Cell 59: 591
17. Collier DN, Bankaitis VA, Weiss JB, Bassford Jr PJ (1988) Cell 53: 273
18. Goldberg AL, Swamy KHS, Chung CH, Larimore FS (1981) Meth Enzym 80: 680
19. Georgiou G, Shuler ML, Wilson DB (1988) Biotechnol Bioeng 32: 741
20. Axe DD, Bailey JE (1987) Biotechnol Lett 9: 83
21. Stouthamer AH (1973) Antonie van Leeuwenhoek 39: 545
22. Stouthamer AH (1973) Biochem Biophys Acta 301: 53
23. Da Silva NA, Bailey JE (1986) Biotechnol Bioeng 28: 741
24. Seo J-H, Bailey JE (1985) Biotechnol Bioeng 27: 1668
25. Bennetzen A, Hall B (1982) J Biol Chem 257: 3026
26. Betenbaugh MJ, Dhurjati P (1990) Ann N Y Acad Sci 589: 111
27. Siegel R, Ryu DDY (1985) Biotechnol Bioeng 27: 28
28. Khosla C, Bailey JE (1988) Nature 331: 633
29. Khosla C, Curtis J, DeModena J, Rinas U, Bailey JE (1990) Bio/Technology 8: 849
30. Mitraki A, King J (1989) Bio/Technology 7: 690
31. Hart RA, Rinas U, Bailey JE (1990) J Biol Chem 265: 12728
32. Chen W, Bailey JE, Lee SB (1991) Biotechnol Bioeng 38: 679
33. Nielsen J, Pedersen AG, Strudsholm K, Villadsen J (1991) Biotechnol Bioeng 37: 802
34. Bentley WE, Kompala D (1989) Biotechnol Bioeng 33: 49
35. Peretti SW, Bailey JE (1986) Biotechnol Bioeng 28: 1672
36. Domach MM, Leung SK, Hahn RE, Cocks GG, Shuler ML (1984) Biotechnol Bioeng 26: 203
37. Imanaka T, Aiba S (1981) Ann N Y Acad Sci 389: 1
38. Ollis DF (1982) Phil Trans Roy Soc London B297: 617
39. Lee SB, Bailey JE (1985) Biotechnol Bioeng 27: 1699
40. Seo J-H, Bailey JE (1985) Biotechnol Bioeng 27: 156

Parameters Influencing the Productivity of Recombinant *E. coli* Cultivations

K. Friehs[1] and K. F. Reardon[2]
[1] Technische Fakultät, AG Fermentationstechnik, Universität Bielefeld, D-4800 Bielefeld, Germany
[2] Department of Agricultural and Chemical Engineering, Colorado State University, Fort Collins, Colorado 80523, USA

Dedicated to Professor Dr. Karl Schügerl on the occasion of his 65th birthday

Advances in Biochemical Engineering
Biotechnology, Vol. 48
Managing Editor: A. Fiechter

In the past 10 to 15 years, many of the promises of microbial genetic engineering have been realized: the use of recombinant *Escherichia coli* has moved from the laboratory to the production facility, and the manufacture of therapeutic recombinant proteins such as human growth hormone and interleukins is a rapidly growing industry.

Along with this progress, however, have come new problems to solve: bioreactor operators have discovered that large-scale cultivations of plasmid-containing bacteria do not behave in exactly the same way as those of plasmid-free cells, plasmid stability has been recognized as a major hurdle, and the protein product might not be present in a soluble form but rather as intracellular granules that resist solubilization. These and other difficulties represent a new generation of challenges for genetic engineering.

However, genetic engineering can do more than solve these problems. Molecular biological techniques also have the ability to create new opportunities: to produce new compounds, to use cheaper substrates, to facilitate downstream processing, and to optimize production in new ways.

The productivity of a cultivation can generally be expressed as the product of the cell density and the specific biological activity. Both of these parameters are influenced by a variety of factors. For recombinant cultivations, though, the level of biological activity, a reflection of the plasmid copy number and expression efficiency, is the more interesting and important consideration and will therefore be given more attention in our review. In this contribution, our general goal is to discuss the factors that influence the productivity of recombinant *E. coli* cultivations, covering

— parameters relating to DNA;
— parameters relating to protein synthesis;
— parameters relating to proteins; and
— parameters relating to downstream processing.

The object is not to tell the reader how to choose the perfect plasmid, host, and cultivation conditions, but to make known the many variables involved in designing a recombinant process and to point out recent and potential advances made possible by genetic engineering. The discussion focuses on the production of a protein, but many of the same concepts apply to other cultivations of recombinant *E. coli*, including cases in which the desired product is not a protein or the cells have been designed for a special metabolic capability such as pollutant biodegradation.

1 Parameters Relating to DNA

1.1 Plasmid Copy Number

It is often true that an increase in the number of copies of a gene (the "gene dosage") will result in an increase in the production of that gene's product protein (the "gene dosage effect"). While this is not always the case – regulatory mechanisms or a saturation effect may impose an upper limit [1] – increasing the copy number of a plasmid is a common method to enhance the productivity of a cultivation.

1.1.1 Increased Copy Number by Plasmid Design

Since the copy number is closely related to plasmid replication, it has been possible to design higher copy number plasmids by modifying replication functions. Examples of this include:

— *cop* and *rom* mutations (in ColE1 derivatives)
— Multiple tandem gene repeats
— Runaway replication plasmids
— Dual low/high copy number origins of replication

The first two of these provide a high plasmid copy number throughout a cultivation. The *cop* and *rom* mutations in ColE1-type plasmids influence the regulation of plasmid replication [2, 3]. These types of mutations have been used to produce ColE1 derivatives maintained at a level of 500 copies per cell [4]. Similar opportunities for exploiting mutations in replication regulation are available for plasmids such as R1, R6K, and pSC101 [5, 6, 7]. The use of tandem gene copies in a plasmid has also been successful; for example, *E. coli* that contained a plasmid with four copies of the chloramphenicol acetyltransferase gene produced four times as much product as cells harboring a plasmid with a single gene copy [8]. However, tandem gene copies are prone to homologous recombination.

Since continuous maintenance of plasmids at high copy numbers is a burden to the host cell, resulting in significantly reduced growth rates as well as increased plasmid instability, plasmid designs allowing one to increase the copy number at an appropriate point in a cultivation are often desirable. One such design is a temperature-sensitive "runaway" replication mutant, with which the copy number of a plasmid is low at 25 °C but rapidly increases when the temperature is shifted to 37 °C [9]. A different method involves the use of both high and low copy number origins of replication on the same plasmid. Here, the idea is to insert a controllable promoter in front of the replication primer of the high copy number plasmid. An example of this is pMG411, which is maintained at 4 copies per cell at 30 °C and at 140 per cell when the culture temperature is increased to 42 °C [10].

1.1.2 Influence of Cultivation Conditions

Cultivation parameters also play a role in determining the copy number; thus, it is also possible to change the copy number of a plasmid by manipulating the cultivation conditions. The effects vary with the plasmid being used, but some trends are evident.

Seo and Bailey [11] utilized different media to change the growth rate (and perhaps other influential parameters) in batch cultures, and observed that the copy number increased with decreasing growth rate. The influences of nutrient limitation in continuous culture have also been studied. When a minimal medium was used in a chemostat with glucose as the limiting substrate, Jones et al. [12] noted significant decreases in copy number after one week of operation.

When using runaway replication plasmids, the culture conditions have a high impact on the achievable copy number. For example, the influence of substrate feeding during runaway induction of pOU140 is shown in Fig. 1 [13]. Clearly, the addition of the carbon source, lactose, led to a significantly greater number of plasmid copies than in the case in which no substrate was added.

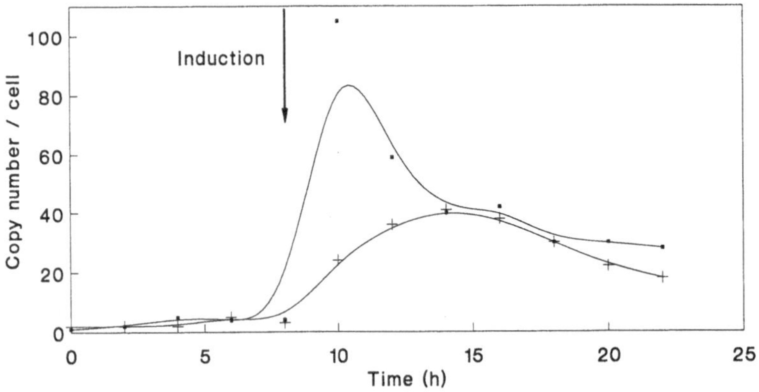

Fig. 1. Runaway replication with addition of lactose medium at point of induction. (—■— with addition; —+— without addition)

1.2 Plasmid Stability

One of the most important issues affecting the productivity of recombinant *E. coli* cultivations is the maintenance of plasmids within the host cells. Plasmid instability is of two types: segregational (plasmid loss resulting from defective partitioning during cell division) and structural (undesired plasmid modifications resulting from insertion, deletion, or rearrangement of DNA).

Factors influencing both types of instability and strategies for overcoming these problems will be discussed in the following sections.

1.2.1 Segregational Genetic Stability

Although some wild type, low copy number plasmids exhibit high stability, most plasmids of industrial interest (high copy number and expressing heterologous proteins) are lost at frequencies of 10^{-2} to 10^{-5} per cell per generation [14]. A recombinant host-vector system with an appreciable level of segregational plasmid instability will be less productive than desired for two reasons:

- the lowered average copy number will generally result in lower specific productivity; and
- plasmid-free cells which eventually emerge have a higher specific growth rate (since they are no longer metabolically burdened by the plasmids) and will thus reach a higher concentration than the plasmid-containing cells.

In a model developed by Imanaka and Aiba [15], cells were grouped as either plasmid-containing or plasmid-free (i.e. copy number was not considered). Two parameters were used to determine the fraction F of plasmid-containing cells remaining after N generations: the frequency of plasmid loss (formation of plasmid-free cells) and the ratio of the growth rates of plasmid-free to plasmid-containing cells. The value of F was much more sensitive to the growth rate ratio than to the loss frequency.

Strategies for overcoming segregational plasmid instability are based on the alteration of one of these key parameters. Several types of approaches can be identified. These have been classified as either selective (based on eliminating plasmid-free cells) vs non-selective [16], or cellular/molecular vs bioprocess methods [17]. In the following paragraphs, methods for preventing or overcoming plasmid instability problems will be discussed according to the level at which they are implemented, i.e.

- plasmid construction
- plasmid copy number
- cultivation conditions
- bioreactor configuration

Influences of plasmid construction. The composition of a plasmid influences its stability and thus provides the basis for the most common approaches for enhancing plasmid maintenance. Perhaps the best known of these is the inclusion of an antibiotic-resistance gene on the plasmid combined with the addition of that antibiotic to the medium. Although this method is simple to implement, it has a number of disadvantages on an industrial scale, including the cost of the antibiotic and the need to separate the antibiotic from the desired product.

In another selective approach, a gene essential to the host cell (usually a mutant) is included in the plasmid. Examples of this are the *serB* gene (for serine production) in a *serB*$^-$ host [18], the *valS* gene (encoding valyl tRNA-synthetase) in a *valS*^{ts} host [19], and the *ssb* gene (for the SSB protein) in a *ssb*$^-$ host [20]. This method has been very successful; for instance, plasmid stability was maintained for more than 200 generations in the *valS* system [19]. However, problems can arise when the mutated host cell is unable to grow rapidly or when spontaneous reverse mutations occur.

Selection for plasmid-containing cells can also be accomplished by killing plasmid-free cells. In one approach, the *parB* locus of plasmid R1 is incorporated into the plasmid. This sequence includes two genes: *hok*, which produces a potent bacteriocin, and *sok*, which forms an mRNA product that prevents the translation of *hok* mRNA. Since the *sok* mRNA has a much shorter lifetime than *hok* mRNA, cells that lose the plasmid after division are killed by the bacteriocin [21]. An alternative method of killing plasmid-free cells is to include the bacteriophage λ repressor on the plasmid and to infect the host cells with the phage. Any plasmid-free cells that appear are then lysed [22].

The use of adjustable copy number (runaway replication) plasmids is a different type of method that allows both good plasmid stability and high productivity. Plasmid stability is maintained during an initial low copy number growth phase due to a favorable ratio of growth rates between plasmid-containing and plasmid-free cells. Following the growth phase, the copy number is increased (e.g. by a temperature shift) and a period of high rates of product formation occurs [9]. This approach can be applied to continuous cultivation by utilizing a two-stage chemostat [23].

Most of the preceding methods are similar in that plasmid-containing cells are given a growth advantage over plasmid-free cells without inherently increasing the segregational plasmid stability. Another problem with selective techniques is that they do nothing to maintain plasmid copy number but rather require only that one plasmid per cell be present.

Non-selective, or genetic, approaches are more promising for ensuring that high copy numbers are stably maintained. An example of these methods is the inclusion of strong terminator sequences on the plasmid to prevent stability problems posed by the strong promoters that are frequently used [24].

The *par* locus of pSC101 is responsible for stable maintenance of that plasmid. When *par* has been cloned into other plasmids, their loss frequencies have decreased [25, 26]. Meacock and Cohen [26] determined that the *par* locus effectively increases the stability of low copy number plasmids but had little impact on high copy number vectors. It should also be noted that insertion of the *par* sequence can reduce the copy number [27].

Another example of a genetic method for enhanced plasmid stability is the incorporation of the *cer* locus of the plasmid ColE1. When plasmids containing *cer* are grown in a xer^{+} (genes for site-specific recombination) host, multimerization of plasmids is reduced, leading to increased plasmid stability [14].

Influences of plasmid copy number. The number of plasmids per cell has an influence on the segregational plasmid stability of the culture. Although low copy number plasmids with an active partition mechanism (e.g. *par*) are usually very stable, high copy numbers generally result in greater stability when random partitioning occurs at cell division [2]. However, this stability advantage of a high copy number system can be offset by the decreased growth rate associated with such cultures, so that the generation of even a few plasmid-free cells can lead to domination by the faster-growing segregant.

Influences of cultivation conditions. The growth environment of a recombinant cell can have a significant effect on the segregational stability of the plasmids it carries. Variations of plasmid stability with differences in growth rate, medium composition, dissolved oxygen concentration, and temperature have been observed; however, it has been difficult to find real trends in most cases.

Plasmid instability has generally been observed to increase with decreasing growth rate, primarily because the relative growth rate advantage of plasmid-free cells over those containing plasmids is decreased under these conditions [23].

The influences of medium composition are less clear, due in part to differences between host strains. Nutrient effects are greater in chemostat cultivations, where at least one nutrient is limiting, than in batch cultures [28]. Although exceptions have been reported, continuous cultures limited by phosphate and magnesium generally exhibit high plasmid instability [12]. Instability under glucose limitation is often lower [29, 30], and nitrogen limitation does not appear to affect plasmid maintenance [29]. Plasmid instability can often be lessened with the addition of complex nutrients like casamino acids [31].

Limiting levels of dissolved oxygen have a detrimental effect on plasmid stability; both short-term (oxygen shock) and long-term oxygen limitations have been shown to increase the rate of plasmid loss [32, 33]. Finally, plasmid stability is generally found to decrease with increased temperature [34].

Bioreactor configuration. Several novel bioreactor designs have been proposed to increase the plasmid stability of a culture. Such alternatives to traditional batch or chemostat cultivations include two-stage chemostats (e.g. for use with runaway replication vectors [23]) and cycling of growth conditions between different dilution rates [35], substrate concentrations [36], and temperatures [37]. A special cell-recycle reactor has been used to maintain high levels of plasmid-containing cells by taking advantage of a flocculation sequence on the plasmid [38]. Cell immobilization has also been shown to increase the plasmid stability of a culture [39].

1.2.2 Structural Genetic Stability

Structural plasmid instability can be difficult to detect in a cultivation, since the growth rate and marker phenotype are the same as those of the desired cells. Several studies have shown that structural instability is affected by the cell's environment; for example, Godwin and Slater found different types of structural changes in glucose- and phosphate-limited chemostats [40].

2 Parameters Relating to Protein Synthesis

2.1 Promoters

The synthesis of a protein starts with the promoter. The initiation of transcription is a rate-limiting process for mRNA synthesis. Comparisons of more than 100 promoters of *E. coli* have shown that there are two regions of conserved DNA sequences [41, 42]. These regions, located 10 and 35 base pairs upstream from the transcription initiation site, strongly influence the strength of a promoter, which in turn determines the rate of transcription initiation. During the development of the molecular biology of *E. coli*, many different promoters were investigated; some of them are currently used in biotechnology and others may be useful after further study.

2.1.1 Promoter Strength

Due to the importance of the promoter strength on the productivity of a recombinant cultivation, genetic engineering is widely used to enhance the strength. The sequence of well known promoters such as *lacUV5* have been changed and the effects on promoter strength examined [43]. New sequences are often tested in order to find especially strong promoters like λP_L and λP_R from the bacteriophage lambda.

While it is important to use strong promoters in the production of recombinant proteins, regulation of those promoters is essential since constitutive overproduction of heterologous proteins leads to decreases in growth rate, plasmid stability, and culture viability. Some promoters are regulated by the interaction of a repressor protein with the operator (a region downstream from the promoter). The most well known operators are those from the *lac* operon and from bacteriophage λ. An overview of regulated promoters in *E. coli* is presented in Table 1.

2.1.2 Induction

A major difference between typical bacterial cultivations and those involving recombinant *E. coli* is the technique of separating growth and production phases. This method takes advantage of regulated promoters to achieve high cell densities in the first phase (while the promoter is "off" and the metabolic burden on the host cell is slight) and then high rates of heterologous protein production in the second phase (following induction to turn the promoter "on").

For industrial bioprocesses, low-cost induction systems are desirable. This is still a problem; for example, the widely used *lac* promoter system is induced with IPTG, a relatively expensive compound. Another common induction technique

Table 1. Regulated promoters in *E. coli*

Induction	Promoter	Operator	Ref.
Temperature shift	λ_{PL}	λ	44
	λ_{PR}	λ	45
	λ tandem	λ	46
IPTG	lac	lac	47
	tac	lac	48
	lac mutations	lac	49
	lpp trp tandem	lac	50
	mac	lac	51
	rac	lac	52
	rrnBP2	lac	53
	Synthetic consensus	lac	54
	T7 gene 10	lac	55
IAA	trp	trp	56
	lpp trp tandem	trp	50
Arsenite	ars	ars	57
Dissolved oxygen	trp	trp	58
	vgb	vgb	59
CO_2 limitation	?	?	60
Fe limitation	?	?	60
Mg limitation	?	?	60
NO_3^{2+} limitation	?	?	61
PO_4^{3+} limitation	ugp	ugp	62
	psi	network	63
N limitation	glnHP2	glnHP2	64
pH shift	alx	alx	65
	λ_{PL}	?	66
Osmotic pressure	bet	?	67
	otsA	?	67
	otsB	?	67
	proU	proU	68
	treA	?	67
Redox potential shift	?	?	69
Succinate	tna	tna	70
Deoxyribose phosphate	deo P1P2	?	71
SOS response	λ_{PL}	?	72
Tryptophan limitation	trp	trp	73

used with promoters like λP_L is a temperature shift, which requires large amounts of energy and may also lead to the undesired formation of stress response proteins. Induction by a pH shift or by addition of an inexpensive inducer molecule would be of great biotechnological interest.

2.2 Terminators

The first step of protein biosynthesis (mRNA formation) ends with the termination of DNA transcription, when the RNA polymerase reaches the terminator sequence. Without correct termination, the mRNA would form the wrong proteins, leading to an overall decrease in productivity. The use of strong promoters requires strong terminators to prevent transcription readthrough. A variety of transcription terminators have been utilized, including the bacteriophage fd terminator [74], and some are commercially available (e.g. the *trpA* terminator) [54].

2.3 mRNA

The first product of protein biosynthesis is mRNA, which is then translated to form the protein of interest. In addition to structural factors of the mRNA that influence its translation, the amount of translatable mRNA has a direct impact on the overall productivity. The amount of mRNA in a cell is a function of the rate of transcription (discussed above) and the rate of mRNA degradation. This degradation rate depends on the presence of RNase recognition sequences, especially those for 3′-exonucleases. Chan et al. have tried to increase the halflife of mRNA by using non-essential regions of an intron [75]. This research led to the development of the cloning vector pKTN-CAT, which increased the production of chloramphenicol acetyltransferase three- to seven-fold by stabilizing the mRNA molecules [76].

A second method to increase the lifetime of mRNA involves the product of the *ams* (altered mRNA stability) gene. Strains that carry the temperature sensitive *ams-1* mutation have longer mRNA halflives at the non-permissive temperature [77].

Another possibility for enhancing the halflife of mRNA is to use RNase-deficient mutants [78]. However, such mutants are often difficult to cultivate.

2.4 Ribosomal Binding Sites

The translation of mRNA, the second step in protein biosynthesis, starts with the binding of the ribosomes at specific binding sites (RBS) on the mRNA molecule. Weak binding sites lead to a low level of expression. In many systems, the expression of foreign genes utilizes the native RBS. An increase in the expression level can be achieved by replacing this natural RBS with an altered, more efficient sequence [79]. For example, Olsen et al. were able to enhance the expression of bovine growth hormone in *E. coli* by enriching the sequence flanking the RBS with A and T nucleotides [80]. The distance between the RBS and the AUG start codon also has an impact on the rate of initiation of translation [81]. A more complete understanding of the influence of the secondary structure of the RBS on the rate of translation should provide further opportunities to optimize foreign gene expression [82].

2.5 *Stop Codons*

As in the case of transcription, translation also requires an efficient termination sequence. These translation stop signals have been added to commonly used vectors like pUC12 (forming pUC12-STOP) [83]. This vector was constructed by inserting a DNA linker with TAA translational stop codons in all three reading frames. Other terminators, such as the Universal Translation Terminator [54], are commercially available.

2.6 *Codon Usage*

When using synthetic genes, the amino acid sequence of a protein is used to develop a DNA sequence. Due to the nature of the genetic code, a choice of codons is usually available. It has been shown that the use of particular codons can play a role in gene expression. For example, highly expressed genes in several species show a bias for certain synonymous codons [84]. Differences in codon usage can also influence mRNA lifetimes.

3 Parameters Relating to Proteins

3.1 *Proteolysis*

Heterologous proteins produced by *E. coli* are usually subject to attack by a variety of proteases. *E. coli* cells produce cytoplasmic, membrane-bound, and periplasmic endoproteases. Relatively little is known about their induction, substrate sequences, or kinetics.

Several methods have been devised to minimize proteolytic activity on the cloned gene product. These include the use of low protease hosts, inhibition of proteases, disguising the desired product by forming a fusion protein, excreting or secreting the product to a "safer" location, and overwhelming the proteolytic enzymes with a high rate of product formation. Modification of cultivation conditions can also reduce protease levels; for example, some proteases are synthesized in response to low dissolved oxygen or glucose conditions.

3.1.1 *Protease Deficient Strains*

The levels of protease activity vary among *E. coli* strains, and this may represent one criterion for host strain selection. More directed efforts have focused on mutants deficient in the production of one or more of the known proteases. The best known examples are the *lon*$^-$ mutants, which cannot form the cytoplasmic protease La [85]. Although cloned protein accumulation can be higher in *lon*$^-$ hosts [86], removal of one protease does not eliminate proteolytic activity. Drawbacks to the use of protease mutant hosts include their altered physiology; for example, *lon*$^-$ strains are UV sensitive and some exhibit a mucoid phenotype that makes cultivation problematic.

3.1.2 Inhibition of Proteases

Although it is more difficult to inhibit intracellular proteases than those present in the cultivation medium, at least one such effort has been successful. In this case, the *pin* sequence (encoding a protease inhibitor produced during bacteriophage T4 infection) was cloned onto a plasmid, effectively reducing the degradation of the desired product [87].

3.1.3 Fusion Proteins

It is often possible to protect a heterologous protein from proteolysis by producing it as a hybrid product with a native protein such as β-lactamase [88] or special foreign proteins like ubiquitin [89]. The use of fused sequences can also enhance translation initiation, and the hybrid products can be designed to facilitate purification. However, the cleavage of the fused molecule to obtain the desired product requires additional processing and may be difficult. Also, expression of a fused protein may be lower than expected.

3.1.4 Protein Export

Another method to reduce proteolysis of the product protein is to engineer its excretion from the cytoplasm into the periplasmic space or its secretion into the medium. Transport into the periplasmic space can be accomplished by fusing leader peptides for periplasmic or outer membrane proteins to the product protein. Although the process of directed transport is not well understood and selection of the best leader sequence is done by trial and error, this approach has proven successful in many cases (e.g. [90]). Additional aspects of protein secretion and excretion are presented in Sect. 4.2.

3.1.5 Rapid Product Formation

Finally, producing the cloned gene product at high rates so as to saturate the proteolytic enzymes is an effective method of minimizing degradation. This can be accomplished with runaway replication vectors or with plasmids containing inducible promoters. Since this strategy has other benefits (e.g. increased plasmid stability), it is frequently implemented.

Rapid production of heterologous proteins often results in the formation of insoluble inclusion bodies. Although these protein aggregates offer further protection from proteases, recovering active product molecules can be problematic.

3.2 Inclusion Bodies

In *E. coli* (and other microorganisms), the product of a foreign gene frequently appears in the form of proteinaceous aggregates called "inclusion bodies" or "refractile bodies". The desired protein can make up from 40 to 95% of the total protein content of these particles; contaminants include other proteins (especially outer membrane proteins), lipopolysaccharides, and membrane fragments [91, 92]. Inclusion bodies can be found in the cytoplasm or in the periplasmic space [93].

The implications of inclusion body formation are mixed for cultivations of recombinant *E. coli* in which a protein is the desired product. (Of course, when the foreign protein is not the desired product, formation of inclusion bodies is detrimental.) Although total protein productivity may be enhanced, the process economics may be adversely affected. These tradeoffs, as well as the factors influencing inclusion body formation, are discussed in the following paragraphs.

3.2.1 Advantages and Disadvantages of Inclusion Body Formation

The production of inclusion bodies can enhance protein production in recombinant *E. coli* cultivations in two areas. First, protein accumulation often increases, primarily because proteins in an inclusion body are resistant to protease attack. The formation of inclusion bodies can also allow the production of high levels of proteins that are toxic to the host cell when soluble.

Due to their physical characteristics, inclusion bodies also offer the opportunity to improve recovery of the protein product from a cultivation. A high degree of purification can be achieved by cell lysis and simple low-speed centrifugation. Since proteins in inclusion bodies are already denatured, additional purification steps can utilize conditions (e.g. detergents) and techniques (e.g. gel filtration chromatography) that are less effective or inappropriate for active proteins.

However, protein production in the form of inclusion bodies can also have significant disadvantages for the bioprocess. One problem involves the release of endotoxins when cells are disrupted to free the aggregates. This is a concern for the purification of any intracellular protein and can only be avoided by secreting the product into the medium.

The larger difficulty posed by inclusion bodies is the renaturation of the protein to obtain an active product; in addition to the technical problems inherent in this task, the process economics can be negatively affected by the special renaturation steps. Following solubilization of the protein particles with concentrated solutions of urea, organic solvents, and/or chaotropic salts (e.g. guanidine-HCl), refolding of the totally denatured protein is required. At present, determining the best process is a trial-and-error procedure. Typical steps are careful removal of the solubilizing agent and the addition of compounds (e.g. thiols) to control the formation of disulfide bonds. Substances like polyethylene glycol have also been reported to improve folding [94, 95]. The in vitro usage of chaperone proteins has been suggested and is an area of active investigation [96, 97]. In general, however, renaturation is still problematic. Although high yields of renaturated proteins have been reported [98], recovery of activity has been low in some cases, especially for larger proteins. In addition, the kinetics of the refolding process are often slow.

An interesting approach to improve the yield and kinetics of renaturation is suggested by the work of Orsini et al. [95], in which removal of a portion of the prourokinase sequence fortuitously resulted in the doubling of the refolding rate and active product yield. If problematic regions (e.g. cysteine-rich) can be altered without affecting the desired activity, similar process improvements might be possible for other proteins.

3.2.2 Factors Influencing Formation

Although the mechanism of inclusion body formation is not well understood, a number of factors influencing their production have been identified. These include aspects of the protein molecule, the rate of protein synthesis, the cultivation conditions, and the host strain used.

Researchers have sought to find a correlation between the formation of inclusion bodies and various protein characteristics. There does not appear to be a connection between the molecular mass, number of disulfide bonds, or hydrophobicity of the protein product [99]. However, an interesting trend has been observed for proteins consisting of subunits; when all necessary subunits are produced concurrently in a cell, they are typically present in the soluble form. On the other hand, inclusion bodies result when individual subunits are produced separately [99].

The rate of protein expression appears to have a major role in determining whether or not soluble protein is produced. High rates of expression generally yield inclusion bodies [100]. Thus, the use of inducible promoters or runaway replication vectors should be expected to lead to the production of inclusion bodies.

Cultivation conditions also affect the form in which foreign proteins appear. Temperature has a significant influence; higher temperatures result in increased inclusion body formation [101]. This could be not only due to an increased rate of synthesis, but also to the relationships between temperature, the protein folding kinetics, and the growth rate of the cells. Similarly, media compositions and pH values that reduce the growth rate have been observed to lead to lower levels of inclusion body formation [93]. In some cases, special medium components such as metal cofactors or nonmetabolizable sugars may exert direct influences on protein folding [102, 103].

Finally, the choice of host *E. coli* strain has been shown to have an impact on the formation of soluble vs particulate protein [104]. As with other aspects of recombinant cultivations that vary among strains, the basis for this influence is not known.

4 Parameters Relating to Downstream Processing

4.1 Cell Harvest and Cell Disruption

If the product is not secreted into the medium or at least exported into the periplasmic space, the cells must be harvested and disrupted to release the product. This is the case for both inclusion bodies and soluble intracellular proteins.

4.1.1 Cell Harvest

The harvesting of cells following a cultivation is basically a liquid-solid separation. Although this is a very common step in the biotechnology industry, it is by no means a simple operation on a large scale. The most frequently used methods are centrifugation and filtration [105]. For large-scale use, only continuous centrifuges such as the tubular bowl or disc models are practical. Common large-scale filtration methods include rotary vacuum drum filters, plate filter presses, and tangential and microfiltration.

It may be possible to employ genetic engineering approaches to improve this separation. One example is a mutation in the *pil* operon that results in the overproduction of pili and subsequent cell flocculation [38]. This leads to an increase in the sedimentation rate and should translate into more rapid centrifugation as well. A similar method that has been developed for yeast involves the cloning of a cell surface protein responsible for flocculation [106].

4.1.2 Cell Disruption

Both mechanical and non-mechanical methods are used to disrupt cells in large-scale processing [105, 107]. Mechanical methods such as high pressure homogenization and bead mills often result in product losses through inefficient disruption or thermal denaturation of proteins. Non-mechanical methods include those based on physical processes (e.g. sonication) and chemical effects (e.g. organic solvents and enzymatic lysis). Each of these have disadvantages; for example, large-scale use of organic solvents has an associated explosion risk [108] and lytic enzymes are expensive [109].

Due to the importance of cell disruption for recovery of recombinant proteins, several genetic approaches have been proposed recently to improve the process with respect to efficiency, active protein yield, waste chemicals, and cost.

One such possibility is the creation of mutant host strains that have a more permeable outer membrane. For example, the amount of porin in the outer membrane of a strain of *E. coli* K12 was altered to increase the permeability [110]. This led to an increase in the export of substances out of the cell. However, mutations such as these frequently result in unhealthy cells that are unable to grow well.

A more elegant technique is to increase the permeability or disrupt the cells at an appropriate time in the cultivation (analogous to the concept of induced transcription). An example of this involves the *kil* gene of ColE1, the product of which leads to total cell lysis [111]. This gene has been placed under the control of the *lac* promoter and incorporated into a plasmid. Thus, lysis can be induced with IPTG after overexpression of the recombinant protein [112]. A similar system has been developed using the lysis gene E from bacteriophage ϕX174, under control of the λP_L promoter [113]. When this system is used together with the temperature sensitive repressor cI857, cell lysis can be induced via a temperature shift to 42 °C.

4.2 Protein Transport

In many cases, it would be desirable to secrete the recombinant product into the medium. Purification would be simpler than for an intracellular protein since the product would not be contaminated with cytoplasmic components. (It would be necessary to handle larger volumes, but this problem has been lessened by newer chromatographic methods.) In addition, the formation of inclusion bodies could be avoided and the toxic effects of some protein products on the host cell could also be reduced. Protein secretion would also reduce proteolysis, unless exoproteases are produced by the transport system.

Excretion into the periplasmic space of *E. coli* also provides many of these benefits. Proteins can be freed from the periplasmic space with gentle treatments that remove the cell wall and outer membrane.

A significant disadvantage facing protein secretion is that protein folding may be incorrect or may not occur. As discussed in the following section, this is a problem in the production of all recombinant proteins. The investigation of special reactors for promoting folding is underway in several laboratories.

Two important factors influence the transport of proteins: the type of cellular transport system and the nature of the signal (or leader) sequences that allow a protein to use that transport system.

4.2.1 Signal Sequences

Signal sequences are short peptides that allow the protein to be transported. In bacteria, these signal sequences are typically 15 to 30 amino acids long [114, 115] and are located at the N-terminus of proteins, although some, like that of hemolysin, are positioned at the C-terminus. Signal sequences form positively charged heads that are able to pass through the membrane. Between the signal peptide and the preprotein, a cleavage site for specific signal peptidases is located to allow formation of the mature protein after transport.

In addition to the leader sequences, transported proteins often have other sequences that are important for secretion. One type, located inside the protein sequence, aids in the formation of a transport competent structure. The function of other sequences, found in membrane proteins, is to stop transport. Thus, the fusion of a gene to a signal sequence does not guarantee that the product will be transported [116].

Some signal sequences that have been used to transport recombinant proteins are shown in Table 2. Well known signal sequences like *bla* (β-lactamase), *malE* (maltose binding protein), *ompA* (outer membrane protein), and *phoA* (alkaline phosphatase) come from proteins that are exported into the periplasm. Fusion proteins formed with one of these are therefore mainly transported into the periplasm, but some secretion has also been observed. Of special interest are the signal sequences *spA* (*Staphylococcus* protein A) and *malE*, since both can be used as purification tags in affinity chromatography (see Sect. 4.4).

4.2.2 Secretion Systems

Different organisms can be used as a secretion system. Bacterial hosts include *Bacillus subtilis, Staphylococcus aureus, Streptomyces lividans*, and, increasingly, *E. coli*. Important eukaryotic secretion systems are yeasts and cell cultures.

The transport of recombinant proteins by one of these hosts can be accompanied by several difficulties [114, 115]. For example, the secretion system may not be able to transport proteins that are too large or have an unfavorable distribution of charges, the wrong hydrophobicity, or similar structural burdens. Other potential problems include the saturation of export sites, competition for the signal peptidases, or lack of the proteins that support secretion.

Genetic engineering can be used to overcome some of the disadvantages of the host. The secretion machinery can be optimized or the permeability of membranes and cell walls can be changed. For *E. coli*, a series of different proteins involved in export to the periplasm are known. Most of these belong to the *sec* product family, and can be coexpressed with the recombinant protein to produce an increase in secretion. Alternatively, the membrane can be changed to

Table 2. Signal sequences and target proteins for transport in *E. coli*; the final protein location is shown as periplasmic space (P) or medium (M)

Signal sequence	Target protein	Location	Ref.
amy	amylase (*B. stearothermophilus*)	M	116
bla	proinsulin (human)	?	117
	IgG (mouse)	?	118
	β-lactamase	M	119
	epidermal growth factor (rat)	P/M	120
	triosephosphatase (chicken)	M	121
malE	gene 5 protein (phage M13)	P	122
	Klenow polymerase	P	123
	nuclease A (*S. aureus*)	P	123
ompA	colony stimulation factor (human)	?	124
	superoxide dismutase (human)	P	125
	α2 interferon	?	126
	antiviral proteins (*Mirabilis*)	M	127
	α-sarcin	P	128
	prokallikrein (human)	P	129
	nuclease A (*S. aureus*)	P	125
ompF	β-endorphin	M	130
phoA	trypsin inhibitor (bovine)	?	131
	epidermal growth factor (human)	?	132
	fusion: β-galactosidase-alk. phosphatase	M/P	133
	α-neo-endorphin	P	134
	fusion: maltose binding protein-β-galactosidase	P	135
	ribonuclease T1	P	136
phoS	growth hormone release factor (human)	P	137
spA	parathyroid hormon (human)	M	138
	insulin-like growth factor (human)	M	139
Ovalbumin	ovalbumin	?	140
Pullulanase	β-lactamase	M	141
Preproinsulin (rat)	proinsulin (rat)	?	142
Enterotoxin LTA	epidermal growth factor (human)	?	143
Synthetic	α2 interferon	P	144
Metalloprotease and helper protein	metalloprotease (with helper protein)	M	145
BRP as helper	insulin-like growth factor (human)	M	131
	cloacin	M	146
hly	hemolysin	M	147

Other abbreviations: amy = amylase, bla = β-lactamase, BRP = bacteriocine release protein, hly = hemolysin, M = medium, malE = maltosebinding protein, ompA = outer membrane protein A, ompF = outer membrane protein F, P = periplasmic space, phoA = alk. phosphatase, phoS = phosphate binding protein, spA = *Staphylococcus aureus* protein A.

yield leaky mutants. Using this approach, alkaline phosphatase [148], β-lactamase [149], and rat proinsulin [150] were secreted into the medium from *E. coli*. The major disadvantage to this method is that leaky mutant hosts are very sensitive to their environment and are difficult to grow. An interesting option is the use of bacteriocin release protein. When expression of this protein was induced, human growth hormone, which had accumulated in the periplasm of *E. coli* due to its fusion to a signal sequence, was secreted into the medium [131].

The export of proteins into the periplasmic space is of increasing interest. The advantages of the periplasmic space, which accounts for 20 to 40% of the total cell volume, include protection against cytoplasmic proteases (Sect. 3.1.4), reduced frequency of inclusion body formation (Sect. 3.2), and the proper environment for correct protein folding (Sect. 4.3). The use of the periplasmic space in industrial scale processes depends on the development of methods to open the cell wall and outer membrane without releasing cytoplasmic substances. At present, the outer membrane is often made more permeable by the addition of chemicals or osmotic shock. Reports of secretion systems in which the outer membrane leakiness is increased by plasmid-encoded products have also been published [151]. Enhanced membrane permeability may also occur fortuitously as a result of high-level expression of a foreign protein.

4.3 Protein Folding

The correct folding of a protein is essential for its biological activity. Although issues of recombinant protein structure and overproduction receive a great amount of attention, relatively little work has been done on protein folding, despite its immense commercial impact and status as an important fundamental question [152, 153]. The factors influencing protein folding are diverse and must be considered when the production of active proteins is desired [154, 155, 156].

The production of recombinant proteins leads to at least two additional problems with regard to protein folding: incorrect folding and the need to renature proteins produced as inclusion bodies.

Some proteins, especially those from eukaryotic sources, are not folded correctly in the oxidative milieu of the bacterial cytoplasm. One way to address this problem is to export the protein into the periplasmic space of *E. coli* (via fusion to a signal sequence). The periplasmic space is a reducing environment and supports correct folding of eukaryotic recombinant proteins.

The need to refold proteins from inclusion bodies can be obviated by preventing their formation. As discussed earlier (Sect. 3.2.2), this can be achieved by using low-expression systems, by changing the cultivation conditions, or by fusion to a signal peptide, among other methods.

4.4 Protein Separation and Purification

A major portion of the production costs in recombinant protein cultivations is due to separation and purification needs. Thus, improvement of these steps via genetic engineering approaches is becoming more common. Purification can be simplified by the secretion of proteins into the medium or excretion into the periplasmic space (Sect. 4.2), but even if this can be achieved, it is still necessary to isolate and purify proteins from a relatively complex mixture.

A good technique for improving protein purification is the use of tags fused to the protein product. The tag (e.g. *Staphylococcus* protein A) is chosen to allow the use of efficient separation techniques such as affinity chromatography. In order to facilitate removal of the tag following purification, a specific cleavage site can be placed between the tag and the desired protein. In Table 3, a number of different tags and their chromatographic ligands are presented.

An example of the power of this method is presented in Fig. 2. In this case, *Staphylococcus* protein A (SpA) was used as a fusion tag for the production of

Table 3. Examples of purification methods using tagged fusion proteins

Interaction	Tag	Ligand	Target protein	Ref.
Immunoaffinity	β-galacto-sidase	anti-β-gal. ab	proline carrier protein	157
Pseudoimmuno-affinity	spA	IgG	EcoRI	158
Substrate affinity	β-galacto-sidase	APTG	DNA binding protein	159
Common binding affinity	malE	starch	gene 5 protein phage M13	160
Metal chelate affinity	his_6	Ni(II)-NTA	DHFR	161
Charge	arg_5	cation exchanger	β-urogastrone	162
Hydrophobic interaction	phe_{11}	phenyl groups	β-galacto-sidase	163
Cysteine thiol interaction	cys_4	thiol groups	galactokinase	163

Abbreviations: ab = antibody, APTG = p-aminophenyl-β-D-thiolgalactoside, DHFR = dihydrofolic acid reductase, IgG = immunoglobulin G, malE = maltose binding protein, NTA = nitrilotriacetic acid, spA = *Staphylococcus aureus* protein A.

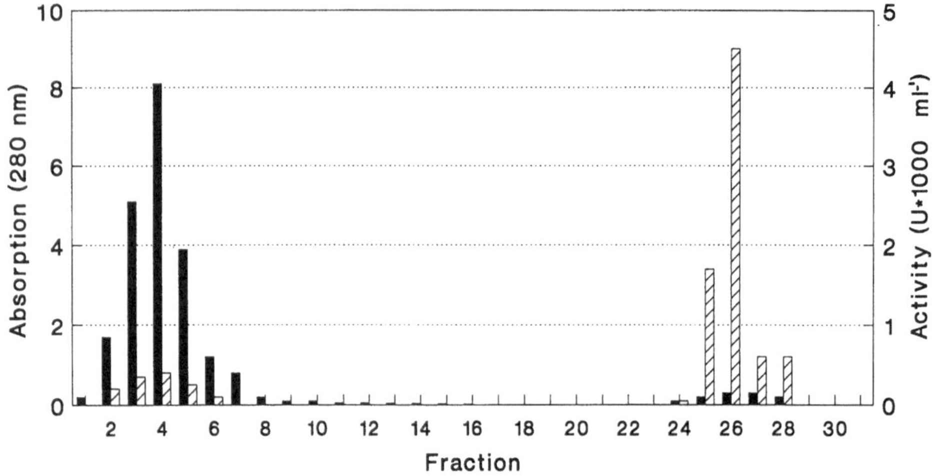

Fig. 2. Affinity chromatography of EcoRI-SpA fusion on IgG ■ Total protein (Abs.) ▨ EcoRI activity

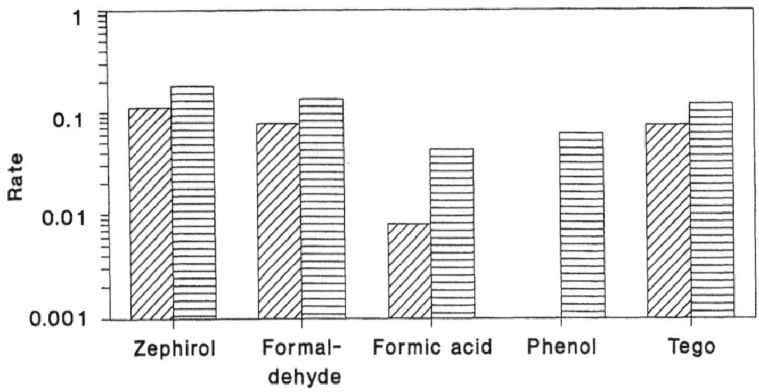

Fig. 3. Inactivation of plasmids by use of disinfectants. Rate of transformation relative to untreated samples. Concentrations: ▨ 1% ▤ 0.1%

the restriction enzyme EcoR1 [158]. The fusion protein could be enriched and purified in one step using affinity chromatography with IgG as the ligand.

4.5 Inactivation of Biological Waste

For safety reasons and for public acceptance, it is necessary to inactivate the biological waste from the production of recombinant proteins. While this is also done for many other cultivations, inactivation of recombinant microorganisms involves not only killing whole cells but also significantly reducing the level of recombinant DNA (such as transformable plasmid DNA).

A series of experiments was performed to determine the efficiency of plasmid inactivation by different methods. The intact biological activity of the plasmids was measured by comparing the transformation rate of isolated plasmid DNA before and after treating the cells. The results of these experiments (shown in Fig. 3) demonstrate that several commonly used compounds, including formaldehyde, formic acid, phenol, Tego, and Zephirol cannot inactivate plasmids when used in normal concentrations [164].

The only truly efficient method to inactivate plasmid DNA with respect to transformation is thermal incubation. This can be accomplished continuously, just as sterilization is.

5 Conclusion

In this review, we have presented a wide variety of factors and techniques that influence the productivity of recombinant *E. coli* cultivations. These parameters exert their effects on many levels, from increased plasmid stability to facilitated product purification.

This information could be used to improve productivity by guiding the design of a vector, the choice of a host strain, or the selection of cultivation conditions and bioreactor type. In addition, the examples provided here suggest additional strategies by which the production of recombinant proteins could be enhanced. The rapid pace of research in this direction should result in many exciting developments in the next few years.

6 References

1. Dennis K, Srienc F, Bailey JE (1985) Biotechnol Bioeng 27:1490
2. Twigg AJ, Sherratt D (1980) Nature (London) 283:216
3. Lacatena RM, Cesareni G (1981) Nature (London) 294:623
4. Tacon WCA, Bonass WA, Jenkins B, Emtage JS (1983) Gene 23:255
5. Terawaki Y, Itoh Y (1985) J Bacteriol 162:72
6. Sherratt D (1986) Control of plasmid maintenance. In: Booth IR, Higgins CF (eds) Regulation of gene expression – 25 years on. Cambridge University Press, Cambridge, p 239 (39th Symp Soc Gen Microbiol)
7. Scott JR (1984) Microbiol Rev 48:1
8. Shibui T, Kamizono M, Teranishi Y (1988) Agric Biol Chem 52:2235
9. Uhlin BE, Molin S, Gustafsson P, Nordstrom K (1979) Gene 6:91
10. Yarranton GT, Wright E, Robinson MK, Humphreys GO (1984) Gene 28:293
11. Seo J-H, Bailey JE (1985) Biotechnol Bioeng 27:1668
12. Jones IM, Primrose SB, Robinson A, Ellwood DC (1980) Mol Gen Genet 180:579
13. Edler C, Friehs K, Schügerl K (1989) DECHEMA Biotechnol Conf 3:549
14. Summers DK, Sherratt DJ (1984) Cell 36:1097
15. Imanaka T, Aiba S (1981) Ann NY Acad Sci 369:1
16. Ensley BD (1986) CRC Crit Rev Biotechnol 4:263
17. Kumar PKR, Maschke H-E, Friehs K, Schügerl K (1991) Trends Biotechnol 9:279
18. Anderson DM, Herrrmann KM, Somerville RL (1983) US Patent No. 4371614
19. Skogman SG, Nilsson J (1984) Gene 31:117
20. Porter RD, Black S, Pannuri S, Carlson A (1990) Bio/Technology 8:47
21. Gerdes K, Rasmussen PB, Molin S (1986) Proc Natl Acad Sci USA 83:3116
22. Rosteck Jr PR, Hershberger CL (1983) Gene 25:29
23. Siegel R, Ryu DYD (1985) Biotechnol Bioeng 27:28
24. Chen J, Morrison DA (1987) Gene 55:179
25. Nishimura N, Taniguchi T, Komatsubara S (1989) J Fermentation Bioeng 67:107
26. Meacock PA, Cohen SN (1980) Cell 20:529
27. Chambers SP (1989) PhD Thesis, University of Warwick
28. Ryan W, Parulekar SJ (1991) Biotechnol Bioeng 37:415
29. Naock D, Roth M, Geutner R, Muller G, Undisz K, Hoffmieir C, Gaspar S (1981) Mol Gen Genet 184:121
30. Jones SA, Melling J (1984) FEMS Microbiol Lett 22:239
31. Brownlie L, Stephenson JR, Cole JA (1990) J Gen Microbiol 136:2471
32. Hopkins DJ, Betenbaugh MJ, Dhurjati P (1987) Biotechnol Bioeng 29:85
33. Caunt P, Impoolsup A, Greenfield PF (1989) Biotechnol Lett. 11:5
34. Wouters JTM, Driehuis FL, Polaczek PJ, Van Oppenraag MLHA, Van Andel JG (1980) Antonie van Leeuwenhoek 37:311
35. Impoolsup A, Caunt P, Greenfield PF (1989) Biotechnol Lett 11:605
36. Stephens ML, Lyberatos G (1985) Biotechnol Bioeng 31:464.
37. Di Pasquantonio VM, Betenbaugh MJ, Dhurjati P (1987) Biotechnol Bioeng 29:513

38. Henry KL, Davis RH, Taylor AL (1990) Biotechnol Prog 6:7
39. Berry F, Sayadi S, Nasri M, Thomas D, Barbotin JN (1990) J Biotechnol 16:199
40. Godwin D, Slater H (1979) J Gen Microbiol 111:201
41. Hawley DK, McClure R (1983) Nuc Acd Res 11:2237
42. Horwitz MSZ, Loeb LA (1990) Prog. Nuc. Acd. Res. Mol. Biol. 38:137
43. Kobayashi M, Nagata K, Ishihama A (1990) Nuc Acd Res 18:7367
44. Reinikainen P, Lähde M, Karp M, Suominen I, Markkanen P, Mäntsälä P (1988) Biotechnol Lett 10:149
45. Botterman J, Höfte H, Zabeau M (1987) J Biotechnol 6:71
46. Elvin CM, Thompson PR, Argall ME, Hendry P, Stamford NPJ, Lilley PE, Dixon NE (1990) Gene 87:123
47. Shapira SK, Chou J, Richaud FV, Casadaban MJ (1983) Gene 25:71
48. Amann E, Brosius J, Ptashne M, (1983) Gene 25:167
49. Makoff AJ, Oxer MD (1991) Nuc Acd Res 19:2417
50. Perez L, Vega J, Chuay C, Menendez A, Ubieta R, Montero M, Padron G, Silva A, Santizo C, Besada V, Herrera L (1990) Appl Microbiol Biotechnol 33:429
51. Vidal-Ingigliardi D, Raibaud O (1985) Nuc Acd Res 13:1163
52. Pistillo, JM, Vishwanatha JK (1990) Biochem Biophys Res Comm 169:1129
53. Lukacsovich T, Orosz A, Baliko G, Venetianer P (1990) J Biotechnol 16:49
54. Pharmacia (1991) Pharmacia LKB Biotechnology, Uppsala, Sweden
55. Deng T, Noel JP, Tsai MD (1990) Gene 93:229
56. Yasukawa K, Saito T (1990) Biotechnol Lett 12:419
57. Owolabi JB, Rosen BP (1990) J Bacteriol 172:2367
58. Flores N, de Anda R, Guerec L, Cruz N, Antonio S, Balbas P, Bolivar F, Valle F (1986) Appl Microbiol Biotechnol 25:267
59. Dikshit KL, Dikshit RP, Webster DA (1990) Nuc Acd Res 18:4149
60. Scanlan DJ, Bloye SA, Mann NH, Hodgson DA, Carr NG (1990) Gene 90:43
61. Berg BL, Stewart V (1990) Genetics 125:691
62. Su TZ, Schweizer H, Oxender DL (1990) Gene 90:129
63. Metcalf WW, Steed PM, Wanner BL (1990) J Bacteriol 172:3191
64. Claverie-Martin F, Magasanik B (1991) Proc Natl Acad Sci USA 88:1631
65. Bingham RJ, Hall KS, Slonczewski JL (1990) J Bacteriol 172:2184
66. Poindexter K, Gayle III RB (1991) Gene 97:125
67. Repoila F, Gutierrez C (1991) Mol Micorbiol 5:747
68. Lucht JM, Bremer E (1991) J Bacteriol 173:801
69. Unden G, Trageser M, Duchene A (1990) Mol Microbiol 4:315
70. Terasawa N, Masayuki I, Uchida Y, Kobayashi M, Kurusu Y, Yukawa H (1991) Appl Microbiol Biotechnol 34:623
71. Fischer M, Fytlovich S, Amit B, Wortzel A, Beck Y (1990) Appl Microbiol Biotechnol 33:424
72. Mott JE, Grant RA, Ho Y, Platt T (1985) Proc Natl Acad Sci USA 82:88
73. Iijima S, Kawai S, Mizutani S, Taniguchi M, Kobayashi T (1987) Appl Microbiol Biotechnol 26:542
74. Stanley KK, Luzio JP (1984) EMBO J 3:1429
75. Chan WKY, Belfort G, Belfort M (1988) Gene 73:295
76. Chan WKY, Belfort G, Belfort M (1991) J Biotechnol 18:225
77. Claverie-Martin F, Diaz-Torres MR, Yancey SD, Kushner SR (1991) J Biol Chem 266:2843
78. Donovan WP, Kushner SR (1986) Proc Natl Acad Sci USA 83:120
79. Schmucker R, Gülland U, Will M, Hillen W (1989) Appl Microbiol Biotechnol 30:509
80. Olsen MK, Rockenbach SK, Curry KA, Tomich C-SC (1989) J Biotechnol 9:179
81. Surek B, Wilhelm M, Hillen W (1991) Appl Microbiol Biotechnol 34:488
82. Spanjaard RA, van Dijk MCM, Turion AJ, van Duin J (1989) Gene 80:345
83. Bröker M, Amann E (1986) Appl Microbiol Biotechnol 23:294
84. Ernst JF (1988) Trends Biotechnol 6:196
85. Maurizi MR, Trisler P, Gottesman S (1985) J Bacteriol 164:1124
86. Boss MA, Kenton JH, Wood CR, Emtage JS (1984) Nucl Acids Res 12:3791
87. Simon LD, Randolph B, Irwin N, Binowski G (1983) Proc Natl Acad Sci USA 80:2059
88. Kadokura H, Yoda K, Mitsunobu I, Yamasaki M (1990) Appl Environ Microbiol 56:2742
89. Wittliff JL, Wenz LL, Dong J, Nawaz Z, Butt TR (1990) J Biol Chem 265:35
90. Quaas R, McKeown Y, Stanssens P, Frank R, Blöcker H, Hahn U (1988) Eur J Biochem 173:617

91. Hartley DL, Kane JF (1988) Biochem Soc Trans 16: 101
92. Langley KE, Berg TF, Strickland TW, Fenton DM, Boone TC, Wypych J (1987) Eur J Biochem 163: 313
93. Georgiou G, Bowden GA (1991) Inclusion body formation and the recovery of aggregated recombinant proteins. In: Prokop A, Bajpai RK, Ho CS (eds) Recombinant DNA technology and applications. McGraw-Hill, New York, p 333
94. Spalding BJ (1991) Bio/Technology 9: 229
95. Orsini G, Brandazza A, Sarmientos P, Molinari A, Lansen J, Cauet G (1991) Eur J Biochem 195: 691
96. Gatenby AA, Viitanen PV, Lorimer GH (1990) Trends Biotechnol 8: 354
97. Horwich AL, Neupert W, Hartl F-U (1990) Trends Biotechnol 8: 126
98. Burton SJ, Quirk AV, Wood PC (1989) Eur J Biochem 179: 379
99. Kane JF, Hartley DL (1988) Trends Biotechnol 6: 95
100. Kopetzki E, Schumacher G, Buckel P (1989) Mol Gen Genet 216: 149
101. Takagi H, Morinaga Y, Tsuchiya M, Ikemura H, Inouye M (1988) Bio/Technology 6: 948
102. Claassen LA, Ahn B, Koo H-S, Grossman L (1991) J Biol Chem 266: 11380
103. Bowden GA, Georgiou G (1988) Biotechnol Prog 4: 97
104. Kenealy WR, Gray JE, Ivanoff LA, Tribe DE, Reed DL, Korant BD, Petteway SR (1987) Dev Ind Microbiol 28: 45
105. Lee S-M (1989) J Biotechnol 11: 103
106. Mota MJ, Teixeira JA (1990) Current Microbiol 20: 209
107. Harrison STL (1991) Biotechnol Adv 9: 217
108. Kula MR, Schütte H (1987) Biotechnol Prog 3: 31
109. Scawen MD, Hammond PM, Sherwood RF, Atkinson T (1990) Biochem Soc Trans 18: 231
110. Sampson BA, Benson A (1987) J Ind Microbiol 1: 335
111. Altieri M, Suit JL, Fan M-LJ, Luria SE (1986) J Bacteriol 168: 648
112. Luria SE, Suit JL, Jackson JA (1991) US Patent 4948735
113. Dabora RL, Eberiel DT, Cooney CL (1989) Biotechnol Lett 11: 845
114. Rasched I, Oberer-Bley E (1990) Chem Lab Biotech 41: 36
115. Hsiung MH, Becker GW (1988) Biotechnol Genet Eng Rev 6: 43
116. Lee C, Li P, Inouye H, Brickman ER, Beckwith J (1989) J Bacteriol 17: 4609
117. Alexander P, Oriel PJ, Glassner DA, Grulke EA (1989) Biotechnol Lett 11: 609
118. Emerick AW, Bertolani BL, Ben-Bassat A, White TJ, Konrad MW (1984) Bio/Technology 2: 165
119. Zemel-Dreasen O, Zamir A (1984) Gene 27: 315
120. Georgiou G, Shuler ML, Wilson DB (1988) Biotech Bioeng 32: 741
121. Ohagi H, Kumakura T, Komoto S, Matsuo Y, Oshiden K, Koide T, Yanaihara C, Yanaihara N (1989) J Biotechnol 10: 151
122. Summers RG, Knowles JR (1989) J Biol Chem 264: 20074
123. Blondel A, Bedouelle H (1990) Eur J Biochem 193: 325
124. Bedouelle H, Duplay P (1988) Eur J Biochem 171: 541
125. Libby RT, Braedt G, Kronheim SR, March CJ, Urdal DL, Chiaverotti TA, Tushinski RJ, Mochizuki DJ, Hopp TP, Cosman D (1987) DNA 6: 221
126. Takahara M, Sagai H, Inouye S, Inouye M (1988) Bio/Technology 6: 195
127. Barbero JOL, Buesa JM, Penalva MA, Perez-Aranda A, Garcia JL (1986) J Biotechnol 4: 255
128. Habuka N, Akiyama KJ, Hideaki T, Miyano M, Matsumoto T, Noma M (1990) J Biol Chem 265: 10988
129. Henze P-PC, Hahn U, Erdmann VA, Ulbrich N (1990) Eur J Biochem 192: 127
130. Shibui T, Matsui R, Uchida-Kamizono M, Okazaki H, Kondo J, Nagahari K, Nakanishi S, Teranishi Y (1989) Appl Microbiol Biotechnol 31: 253
131. Sherwood R (1991) Trends Biotechnol 9: 1
132. Marks CB, Vasser M, Ng P, Henzel W, Anderson S (1986) J Biol Chem 261: 7115
133. Oka T, Sumi S, Fuwa T, Yoda K, Yamasaki M, Tamura G, Miyake T (1987) Agricult Biolog Chem 5: 1099
134. Ohsuye K, Nomura M, Tanaka S, Kubota I, Nakazato H, Shinagawa H, Nakata A, Noguchi T (1983) Nuc Acd Res 11: 1283
135. Guan C, Li P, Riggs PD, Inouye H (1988) Gene 67: 21
136. Fujimura T, Tanaka T, Kanako O, Morioka H, Uesugi S, Ikehara M, Nishikawa S (1990) FEBS 269: 71

137. Anba J, Baty D, Lloubes R, Pages J-M, Joseph-Liauzum E, Shire D, Roskam W, Lazdunski C (1987) Gene 53:219
138. Hogset A, Blinsmo OR, Saether O, Gautvik VT, Holmgren E, Hartmanis M, Josephson S, Gabrielsen OS, Gordelaze JO, Alestrom P, Gautvik KM (1990) J Biol Chem 265:7338
139. Wadenstein H, Ekebacke A, Hammarberg B, Holmgren E, Kalderen C, Tally M, Moks T, Uhlen M, Josephson S, Hartmanis M (1991) Biotechnol Appl Biochem 13:412
140. Fraser TH, Bruce TJ (1987) Proc Natl Acad Sci USA 75:5936
141. Kornacker MG, Pugsley AP (1990) Molec Microbiol 4:1101
142. Talmadge K, Brosius J, Gilbert W (1981) Nature 294:176
143. Morioka-Fujimoto K, Marumoto R, Fukuda T (1991) J Biol Chem 266:1728
144. Legoux R, Leptlatois P, Joseph-Liauzun, E (1991) US Patent 4945047
145. Letoffe S, Delepelaire P, Wandersman C (1991) J Bacteriol 173:2160
146. van Putten AJ, de Graaf FK, Oudega B (1987) Proc 4th Europ Con Biotechnol 4:593
147. Mackman N, Nicaud JM, Gray L, Holland IB (1986) Cur Top Microbiol Immunol 125:159
148. Lazzaroni JC, Portalier RC (1982) Eur J Appl Microb Biotechnol 16:146
149. Atlan D, Portalier RC (1984) Eur J Appl Microb Biotechnol 19:5
150. Mosbach K, Birnbaum S, Hardy K, Davies J, Bulow L (1983) Nature (London) 302:543
151. Murakami Y, Furusato Y, Kato C, Habuka N, Kudo T, Horikoshi K (1989) Appl Microbiol Biotechnol 30:619
152. Bialy H (1987) Bio/Technology 5:883
153. Thatcher DR (1990) Biochem Soc Trans 18:234
154. Fischer G, Schmid FX (1990) Biochem 29:2205
155. Dill KA (1990) Biochem 29:7133
156. Jaenicke R (1987) Prog Biophys Molec Biol 49:117
157. Hanada K, Yamamoto I, Anraku Y (1987) J Bio Chem 262:14100
158. Maschke H-E, Hebenbrock K, Friehs K (1990) DECHEMA Biotechnol Conf 4:379
159. Ullman A (1984) Gene 29:27
160. Maina CV, Riggs PD, Grandea AG, Slatko BE, Moran LS, Tagliamonte JA, McReynolds LA, di Guan C (1988) Gene 74:365
161. Hochuli E, Bannwarth W, Döbeli H, Gentz R, Stüber D (1988) Bio/Technology 11:1321
162. Smith JC, Derbyshire RB, Cook E, Dunthorne L, Viney J, Brewer SJ, Sassenfeld HM, Bell LD (1984) Gene 32:321
163. Persson M, Bergstrand MG, Bülow L, Mosbach K (1988) Anal Biochem 172:330/337
164. Stiegelmeier C, Friehs K, Schügerl K (1992) Acta Biotechnol (in press)

Modeling and Control for Anaerobic Wastewater Treatment

Elmar Heinzle[1], Irving J. Dunn and Gerhard B. Ryhiner[2]
Biological Reaction Engineering Group, Chemical Engineering Department, Swiss Federal Institute of Technology, Zürich, Switzerland

Dedicated to Professor Dr. Karl Schügerl on the occasion of his 65th birthday

The recent literature on the modeling and control of anaerobic wastewater treatment processes is reviewed. An example from the author's personal work is used to describe how a dynamic simulation model can be developed from the basis of multi-organism growth kinetics and mass balancing techniques. This included consideration of the organic acid dissociation equilibria important for pH calculation and the thermodynamic influence of hydrogen on the reactions involving propionic and butyric acids. Liquid phase balances were linked to gas phase balances by gas-liquid transfer considerations. It is shown in detail how the model was applied to one and two stage experimental reactors for the design and tuning of controllers. Both conventional PID controllers and adaptive optimizing controllers, employing simple input-output models with an objective function, were tested.

[1] To whom all correspondence should be addressed
[2] Present address: Sulzer Chemtech, 8401 Winterthur, Switzerland

Advances in Biochemical Engineering
Biotechnology, Vol. 48
Managing Editor: A. Fiechter

List of Symbols

Symbol	Unit	Name
A	$[m^2]$	Reactor cross sectional area
a_i, b_i, c	[–]	Model parameter
C	$[mol\ m^{-3}]$	Concentration of inorganic compound
C	$[C\text{-}mol\ m^{-3}]$	Concentration of organic compound
EF	[–]	Equilibrium factor
F	$[m^3\ h^{-1}]$	Liquid flow rate
G	$[mol\ h^{-1}]$	Gas flow rate
$\Delta G^{0'}$	$[kJ\ mol^{-1}]$	Free Gibb's enthalpy
ΔG_f^0	$[kJ\ mol^{-1}]$	Standard free enthalpy of formation
H_i	$[mol\ m^{-3}\ bar^{-1}]$	Henry-coefficient
K_S	$[C\text{-}mol\ m^{-3}]$	Saturation constant
K_I	$[C\text{-}mol\ m^{-3}]$	Inhibition constant
K_A	$[mol\ m^{-3}]$	Acid dissociation constant
K_B	$[mol\ m^{-3}]$	Base dissociation constant
K_W	$[mol^2\ m^{-6}]$	Water dissociation constant
K_r	[*]	Controller gain
k_d	$[h^{-1}]$	Death rate of organisms
k_La	$[h^{-1}]$	Gas liquid phase mass transfer coefficient
N	$[mol\ m^{-3}\ h^{-1}]$	Mass transfer rate
n_i	[mol]	Amount of gas i in reactor gas phase
p	[bar]	Total pressure
PI	[–]	Performance index
R	$\left[\frac{bar\ m^3}{mol\ K}\right]$	Universal gas constant
r	$[mol\ m^{-3}\ h^{-1}]$	Reaction rate
r_P	$[C\text{-}mol\ m^{-3}\ h^{-1}]$	Product formation rate
r_S	$[C\text{-}mol\ m^{-3}\ h^{-1}]$	Substrate consumption rate
r_X	$[C\text{-}mol\ m^{-3}\ h^{-1}]$	Growth rate
T	[K]	Temperature
u_B	$[m\ h^{-1}]$	Bubble rising velocity
u	[*]	Manipulated process input
V	$[m^3]$	Volume
X	$[C\text{-}mol\ m^{-3}]$	Biomass concentration
x_i	[–]	Molar fraction of gas i in bubbles
$Y_{X/S}$	$[C\text{-}mol\ C\text{-}mol^{-1}]$	Biomass yield from substrate S
$Y_{P/S}$	$[C\text{-}mol\ C\text{-}mol^{-1}]$	Product yield from substrate S
y	[*]	Process output

Greek symbols:

α	[*]	Gain adaptive optimizer
β	[*]	Working point optimizer

δ	[mol m^{-3}]	Different quantity in Eq. (14)
ε_g	[–]	Fractional gas hold up
λ_{min}	[*]	Minimum forgetting factor optimizer
μ	[h^{-1}]	Specific growth rate
μ_{max}	[h^{-1}]	Maximal specific growth rate
ν_i	[–]	Stoichiometric coefficient
Σ	[*]	Sensitivity optimizer
τ_{sl}	[h]	Solid retention time

[*]: Units varying depending on application

Indices:

AH	acid
A^-	dissociated AH
Ac	acetic acid
An^-	anion
B	base
Bu	butyric acid
D	differential
g	gas phase
HAc	undissociated acetic acid
HBu	undissociated butyric acid
HPr	undissociated propionic acid
i	refers to component
I	integral
K^+	cation
L	liquid phase
P	product
Pr	propionic acid
S	substrate
sl	solid
titr	titrator base
tot	total
W	water
X	biomass
Z	surplus cations
0	reactor feed stream

Abbreviations:

COD	Chemical oxygen demand
GC	Gas chromatograph(y)
P	Proportional
PI	Proportional-integral
PID	Proportional-integral-differential
VFA	Volatile fatty acids

1 Introduction

Anaerobic processes have obvious advantages for wastewater treatment, especially high strength wastewaters. These are a sludge production decrease (one-tenth that of aerobic treatment), no aeration energy requirement, reduction of odors in a closed system, and methane energy production. Distinct disadvantages remain: methanogenic organisms grow slowly, the stability of anaerobic processes can be upset either by toxic substrates or by overloading, and the process is not completely understood.

Much progress has been made in the understanding of anaerobic processes, from both the microbiological and from the theoretical standpoint. This contribution will attempt to discuss and evaluate recent efforts in the mathematical modeling and in the design of control strategies for anaerobic wastewater treatment. In some respects the two areas are related, since modeling leads to quantitative understanding and control system design benefits from this. Simple control methods are based on simple qualitative ideas, but it will be shown that a model can be useful in simply reducing the time necessary to test the controller. The slow rate of anaerobic processes makes it useful to minimize experimentation and to replace it as far as possible with simulation studies. This combined approach using simulation and experimentation to develop control strategies and to determine control parameters will be emphasized here.

In discussing the model developments it must be kept in mind that in anaerobic wastewater treatment one deals with a multiorganism system, of yet unknown complexity, with respect to the number of reactions and their intermediates. In addition, the spatial configuration of the individual organisms in flocs, granules and biofilms is largely unknown and would itself have a pronounced influence on the overall process, for example due to hydrogen scavenging. Comparing the multiorganism-system to that of a single organism and remembering that no single kinetic model will describe even all the complexities of pure culture dynamics, then it must be a combination of naivety and fearlessness that allows engineers and scientists to attempt such model development.

Modelling must be viewed in the light of its purpose. If the understanding of the process complexities is the goal, then a detailed model may be justified, especially if the many variables are actually measured and compared with the model. Complex models may be very useful for design of state estimators and controllers. For the purpose of model based state estimation, it is necessary to simplify a complex mechanistic model to the point that it can be treated with the available theory. A curve-fitting non-mechanistic model may be another approach possible if on-line measurements are available.

2 Recent Reviews of the Field

Andrews [1] developed a simulation model based on Monod kinetics with substrate inhibition for growth, using a constant yield for consumption and production. Undissociated fatty acid was considered as the substrate for the formation of methane, and the dissociation equilibrium was considered. In addition the bicarbonate equilibrium and a charge balance was used to obtain dissolved CO_2 and pH. The transport rate of CO_2 to the gas phase was considered to obtain the total gas rate. The simulations gave dynamic results in terms of un-ionized acid substrate, bicarbonate, total gas flow, gas composition. Thus it was possible to consider the response of the continuous digester to step changes in substrate feed. This model permitted the simulation of various control strategies, such as recycling of thickened sludge, and the scrubbing of carbon dioxide from the liquid phase using recycle methane. This early work represented the state of the art in anaerobic process modeling, part of which was done on a hybrid analog-digital computer.

One of the leaders in the applied anaerobic treatment field has reviewed the research and development in anaerobic treatment during the last century [34]. From this review the microbial complexities are evident, and it is also clear that much has still to be done regarding basic research and process innovation.

Rozzi et al. [50] has more recently used the Andrews model with some modifications: gas-liquid equilibrium was assumed, and the fatty acid substrate was considered to be acetic acid which reacted with the bicarbonate. Control of pH with the addition of NaOH, Na_2CO_3 and Na_2HCO_3, and of bicarbonate alkalinity with NaOH addition was studied with simulation. Also considered was an early warning on the basis of CO_2 in the off-gas for conventional pH control with the base addition. A later paper [51] considered bicarbonate alkalinity control in more detail and discussed its automatic measurement. This work concluded that bicarbonate alkalinity control is promising. Wiesmann [69] gave an overview with special emphasis on acetic acid consumption and control of pH. Schürbüscher and Wandrey [55] reviewed many aspects of anaerobic digestion including reaction schemes, contribution of various organism groups, modeling, on-line measurement and estimation as well as control.

Barnes and Fitzgerald [5] gave an overview on anaerobic reactions and kinetics with special emphasis on the influence of toxic compounds (metals and organics).

An extended overview on the biology of various anaerobic processes is given in the books edited by Zehnder [70] and by Bélaich et al. [54]. In an earlier book [22] many aspects including modeling are extensively treated. The most important chapters related to the topics of the present paper are on the modeling and its application for start-up and control [58] and on energetics and biomass yields of anaerobic processes [59].

A recent review [41] stated that with the exception of the hydrolysis of solids all other subprocesses of anaerobic digestion have been modeled successfully

Table 1. Rate parameters as reported from the literature by Pavlostathis and Giraldo-Gomez [41]. VSS – volatile suspended solids

Substrate	$k\left(=\frac{\mu_{max}C_S}{K_S Y}\right)$ $\frac{g\,COD}{g\,VSS\,d}$	K_S $\frac{g\,COD}{m^3}$	μ_{max} d^{-1}	Y $\frac{g\,VSS}{g\,COD}$
Butyric acid Propionic acid	6.2–17.1	12–500	0.13–1.2	0.01–0.27
Acetic acid	2.6–26	11–930	0.08–0.7	0.01–0.054
H_2 and CO_2	1.92–90	4.8×10^{-5}–0.6	0.05–4.07	0.017–0.13

with Monod-type kinetics. This includes the conversion of sugars to organic acids, the conversion of acetic acid to methane and the reduction of carbon dioxide by hydrogen to methane. The reasons given for the wide variation of kinetic parameters shown in Table 1, where the widely varying conditions, and possibly inaccurate measuring procedures. Certainly the complexity of anaerobic processes is also reflected in these widely varying kinetic constants.

3 Reactions in Anaerobic Digestion

Anaerobic degradation is a very complex multi-substrate, multi-organism process. Although there are still some mysteries, most mechanistic models consider basically the four steps shown in Fig. 1.

Here the polymer materials (carbohydrates, proteins or lipids) are hydrolyzed to yield the monomer compounds (e.g. amino acids, sugars and fatty acids). In some cases particles are contained in the wastewater. Structured modeling of the digestion of particulate material has been addressed by Bryers [8] and Tschui [62]. Anaerobic hydrolysis of fats and proteins is a complex process, involving various groups of organisms but is quite well understood [37]. Many compounds are not easily degraded to short chain carboxylic acids. These include especially aromatic compounds like lignin-derived materials [9].

In a second acidogenic step these compounds are transformed to organic acids (mainly acetic, propionic and butyric acid) carbon dioxide and hydrogen. In the third acetogenic step organic acids with more than three atoms of carbon per molecule are converted to acetic acid and hydrogen [20]. The last methanogenic steps convert acetic acid to methane and reduce carbon dioxide with hydrogen to methane, a process well investigated but incompletely understood on the biochemical level [63].

In Table 2 the most important reactions are summarized with whey as the starting substrate. Only the elements C, H and O are considered in these reactions.

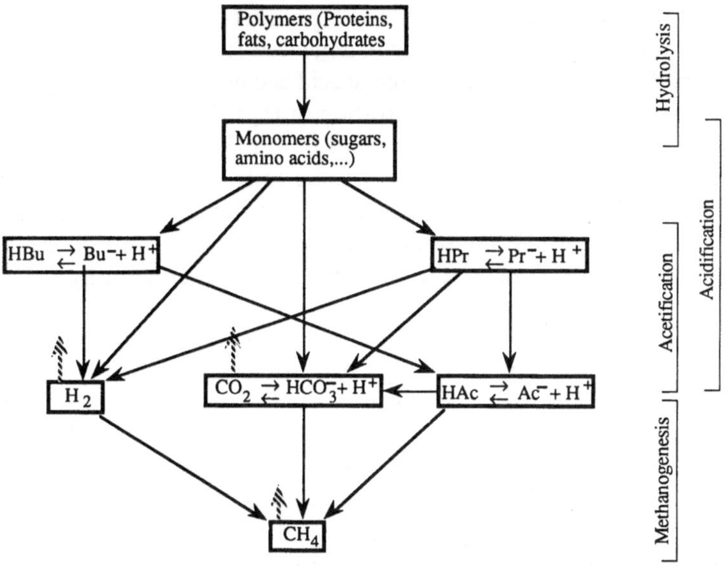

Fig. 1. Reaction scheme of anaerobic degradation. *Dashed arrows* indicate gaseous compounds transferred out of the liquid phase. HPr – propionic acid; Pr^- – propionate; HBu – butyric acid; Bu^- – butyrate; HAc – acetic acid; Ac^- – acetate

Table 2. Most important reactions in the anaerobic degradation of whey. Reaction R1 is given in a C-mol basis (one carbon atom per molecule); ν_i – stoichiometric coefficients. ($\Delta G^{0'}$ for T = 25 °C, pH = 7, p = 1 bar, $x_{H_2O} = 1$, c = 1 mol kg^{-1})

Hydrolysis and acidification of whey			
$CH_{1.85}O_{0.853} + \nu_{H_2O} H_2O \rightarrow \nu_{Bu} CH_2O_{0.5} + \nu_{Pr} CH_2O_{0.667} + \nu_{Ac} CH_2O + \nu_{CO_2} CO_2 + \nu_{H_2} H_2$			(R1)
Acetification			
$CH_3(CH_2)_2COO^- + 2H_2O$	$\rightleftarrows 2\,CH_3COO^- + 2\,H_2 + H^+$	$\Delta G^{0'} = +\ 71.7$ kJ	(R2)
$CH_3CH_2COO^- + 2\,H_2O$	$\rightleftarrows CH_3COO^- + CO_2 + 3\,H_2$	$\Delta G^{0'} = +\ 48.3$ kJ	(R3)
Methanogenesis			
$CH_3COO^- + H^+$	$\rightleftarrows CH_4 + CO_2$	$\Delta G^{0'} = -\ 35.8$ kJ	(R4)
$CO_2 + 4\,H_2$	$\rightleftarrows CH_4 + 2\,H_2O$	$\Delta G^{0'} = -\ 130.7$ kJ	(R5)

Referring to Table 2, two typical process cases may be distinguished.

- If solid matter has to be degraded, then the hydrolysis step may be rate limiting, and overloading will not be important.
- Overloading of the reactor may occur if easily degradable material is to be treated. It is agreed that hydrogen plays an important role in the process stability. Due to the positive free energy of reaction, reactions R2 and R3 in

Table 2 can only proceed to the right, if the products, especially H_2, are continuously removed. Thus if the final methanogenesis steps are inhibited due to toxicity or pH then the hydrogen and acetic acid accumulation will stop the reactions, R2 and R3. Because the concentration of H_2 influences the equilibrium relation to the second (R2) and even third power (R3), it is evident that the hydrogen partial pressure must be kept sufficiently low. This is accomplished by the methanogenic scavengers, whose rates cannot keep up in the event of substrate overloads. H_2 consuming organisms can remove H_2 locally to concentration levels which make the reaction proceed exergonically [11].

According to Thauer [61] and Archer [2] the oxidation of propionic to acetic acid should thermodynamically only be possible at $pH_2 < 10^{-4}$ bar. Experimental measurements of dissolved hydrogen in this range are difficult to obtain. Experiments by Kaspar [28], Kaspar and Wuhrmann [29] and Denac and Dunn [15] applying pure H_2 did not show any significant inhibition. Kaspar [28] and recently Whitmore et al. [68], who actually measured dissolved H_2 down to the 1 µmolar range, pointed out that this may be caused by diffusion-reaction phenomena. Local high consumption rates of H_2 in a biofilm or floc may reduce local H_2 concentration to allow the endergonic reactions R2 and R3 to proceed. The importance of hydrogen as controlling intermediate was extensively reviewed by Harper and Pohland [26] and in recent book [54].

From a series of observations it seems to be evident that disturbances in the methanogenic step, which generally is considered to be the most sensitive one, leads to accumulation of H_2 [2, 26, 35, 67]. Additionally the state of an anaerobic reactor may be characterized by its volatile fatty acid levels, from the CH_4/CO_2 ratio and from the total gas production rate.

The compounds shown in Fig. 1 and Table 2 are not the only ones occurring in anaerobic digestion. Investigating the influence of hydrogen on the thermodynamics of reaction in a reactor continuously fed with ethanol and propionate, it was found that a shock load of feed did not just stop the propionate conversion, as expected from hydrogen considerations, but that it induced the reduction of the propionate to propanol [57].

A new modeling concept was proposed by McCarty and Mosey [36] based on complex population dynamic interactions of "butyric bacteria", which produce butyric acid at low pH and acetic acid normally from glucose, and "propionic bacteria", which produce propionic and acetic acid at neutral pH. The "propionic bacteria" are seen to be mainly *Enterobacteria* that grow well at high concentrations and neutral pH, while the "butyric bacteria" are Clostridia that grow well at low concentrations, producing acetic acid but produce butyric acid at low pH. Also involved are the parallel reactions of *Syntrophomonas*, which converts higher organic acids to propionic and acetic acids, and *Syntrophobacter*, which utilizes only propionic acid. Both rely on the hydrogen consumption of *Methanobacterium* for their thermodynamics. Thus a reactor overload would first cause the acetate and propionate to increase, lowering the pH; this would then cause the formation of butyrate. Hydrogen would quickly accumulate and be consumed, as would acetate. Propionate would fall slowly because only the *Syntrophobacter* can use it. One thing is clear from the new modeling concepts presented in this paper: many mysteries still remain with regard to the microbiology and its kinetic modeling.

Smith and McCarty [57] used a four bacteria-type model for mixed substrates ethanol and propionate. Two types were used for oxidation of the two substrates to acetate and the other two types for methanogenesis. Thus four component balances and four biomass balances were used.

Costello et al. [12] used the Mosey model [39] as a basis but also included lactic acid as an important intermediate. Six groups of bacteria were considered. One group formed acetic, butyric or lactic acid and a second degraded lactic acid into propionic or acetic acid. Two groups were acetogenic and formed acetic acid from butyric and from propionic acid. Methane was formed by two groups, one by hydrogen-consuming bacteria reducing carbon dioxide and another from acetate.

A four organism Mosey-type model was used by Dochain et al. [19] to develop an adaptive control algorithm for the control of hydrogen concentration. A quasi-steady state assumption regarding the glucose concentration allowed relating the accumulation rate for hydrogen to the glucose inflow rate and the hydrogen outflow gas rate. A further assumption allowed neglecting the propionate reaction path. For the simulated results both hydrogen and acetate were assumed to have inhibiting effects on the methanogenic steps and the originally neglected propionate reaction was included. Without control the hydrogen-consuming methanogens would be washed out of the reactor.

Ryhiner [52] used the reaction scheme shown in Fig. 1 and Table 2 including a six organism group model to describe anaerobic digestion of whey.

4 Dynamic Mass Balance Equations

For the purpose of modeling, many reactors used in anaerobic degradation can be considered as well mixed. The basic balance equation for a well mixed liquid phase is

$$\begin{bmatrix}\text{accumulation} \\ \text{of component i}\end{bmatrix} = \text{input} - \text{output} + \text{reaction} + \begin{bmatrix}\text{mass transfer from} \\ \text{the gas to the} \\ \text{liquid phase}\end{bmatrix}$$

$$V_L \frac{dC_{L_i}}{dt} = F_L\,(C_{L_{i0}} - C_{L_i}) + r_i\,V_L + N_i\,V_L \qquad (1)$$

4.1 Biomass Balance

The simplest approach considers biomass to be homogeneously suspended in the reactor. In many reactor systems the biomass residence time is made longer than the hydraulic residence time, either by biomass recycling or by immobilization. The biomass retention can be modeled using a sludge retention time (τ_{sl}) which is always larger than the hydraulic retention time. The resulting biomass balance is:

$$\text{accumulation} = \text{production} - \text{death} - \begin{bmatrix} \text{loss in} \\ \text{effluent} \end{bmatrix}$$

$$\frac{dX_i}{dt} = r_{X_i} - X_i \cdot k_{d_i} - \frac{X_i}{\tau_{sl}} \tag{2}$$

where

$$r_{X_i} = \mu_i X_i \tag{3}$$

where i indicates populations feeding on the various substrates (whey, butyric, propionic and acetic acid, hydrogen). The term $X_i k_{d_i}$ also accounts for maintenance and endogenous reactions.

4.2 *Substrates and Product Balances*

For non-volatile substrates, intermediates and products Eq. (1) simplifies to

$$\begin{bmatrix} \text{accumulation} \\ \text{of component i} \end{bmatrix} = \text{flow} - \text{consumption} + \text{production}$$

$$V_L \frac{dC_{L_i}}{dt} = F_L (C_{L_{io}} - C_{L_i}) + \sum r_{S_i} V_L + \sum r_{P_i} V_L \tag{4}$$

Each substrate may be consumed and produced by several reactions which are taken into account by the summation. The corresponding balances for volatile components (H_2, CO_2 and CH_4) include mass transfer term as shown in Eq. (1).

The relation of substrate consumption is

$$r_{S_i} = \frac{-\mu_i X_i}{Y_{X_i/S_i}} \tag{5}$$

Formation of a product j from substrate i is usually assumed proportional to biomass formation. This is expressed using yield coefficients

$$r_{P_j} = \frac{\mu_i X_i Y_{P_j/S_i}}{Y_{X_i/Si}} \tag{6}$$

5 Stoichiometry

Stoichiometric coefficients for reactions R2 to R5 in Table 2 are well defined. Only the stoichiometric coefficients (v_i) of reaction R1 and corresponding other hydrolysis reactions are not known a priori. Experiments under defined conditions are necessary to determine these coefficients. The six unknown stoichiometric coefficients of the acidification of whey were determined after measurement of acid production and COD reduction in the acidification reactor of the two stage system [52]. In this reactor no methane was produced, and reactions R2 and R3 were assumed not to proceed under these conditions. The experimentally determined values were: $\nu_{Bu} = 0.5$; $\nu_{Pr} = 0.12$; $\nu_{Ac} = 0.15$; $\nu_{CO_2} = 0.23$; $\nu_{H_2} = 0.24$; $\nu_{H_2O} = 0.095$.

More difficult is the estimation of biomass yield coefficients of individual reactions. One successful approach uses ATP balancing. The yield of biomass from ATP ($Y_{X/ATP}$) is quite constant for many organisms [4, 48]. The average value given there is $Y_{X/ATP} \approx 10$ g mol^{-1}. If the production of ATP from metabolic reactions is known, the biomass yield can be estimated in a straightforward manner. Smith and McCarty [57] developed these stoichiometric coefficients from the free energy of the catabolic reactions. Growth was related to energy availability by calculating the difference between the free energy of the catabolic reaction and the energy involved in the conversion of the substrates to pyruvate. It was noted that at least 6 reactions occurred during the experiments, which involved reduced intermediates that were not included in the model. Erickson [21] presented an electron balance and discussed it as a means of estimating biomass yield coefficients in anaerobic processes.

6 Kinetics

The hydrolysis of especially particulate material is a very complex process. Often first order kinetics with respect to particles are applied [62]. In the work of Ryhiner [52] the kinetics of biomass growth are described by simple Monod relationships for whey, butyric acid, propionic acid and hydrogen.

$$\mu_i = \frac{\mu_{max_i} C_{S_i}}{C_{S_i} + K_{S_i}} \tag{7}$$

or with an additional substrate inhibition term (acetic acid)

$$\mu_i = \frac{\mu_{max_i} C_{S_i}}{C_{S_i} + K_{S_i} + C_{S_i}{}^2/K_{I_i}} \tag{8}$$

The acidic form was taken to be the kinetically determining form, as was done in previous work [14, 32, 69]. The concentrations of the actual substrates (undissociated carboxylic acids) are pH dependent. At neutral pH-values, only CO_2-/HCO_3- and in sulfur-containing wastewaters H_2S/S^- have a reasonable buffer capacity. All other acids mainly exist in their salt form.

In the acetogenic step, acetic acid, hydrogen and carbon dioxide are produced from propionic and butyric acid. The thermodynamic limits for these reactions are incorporated by estimating the chemical equilibrium limits for butyric acid and propionic acid using an empirical factor which was varied as a function of the ratio of the time varying equilibrium constant, based on the instantaneous concentrations, to the true value based on $\Delta G^{0'}$. The value of the factor varied as an S-shaped function from 0 to 1 [52]. Another but similar approach was taken by Mather [33], who also used Eq. (7) to describe specific growth rate, but set $\mu_{max} = K\,\Delta G$. Here ΔG was the actual free reaction enthalpy at actual concentration values and K was an empirical constant, which caused μ_{max} to increase monotonously with increasing substrate and decreasing product concentrations. No real saturation value for growth rate at high substrate concentration is obtained.

In a recent review [41] it was stated that with the exception of the hydrolysis of solids all other subprocesses of anaerobic digestion can be modeled successfully with Monod-type kinetics.

Table 3. Anaerobic stoichiometric and kinetic parameters from the literature. (C-molecular weight of biomass: 29.5 g C-mol^{-1})

Substrate	μ_{max} [h^{-1}]	K_s [C-mol m^{-3}]	K_I [C-mol m^{-3}]	$Y_{X,S}$ [C-mol C-mol^{-1}]	k_d [h^{-1}]	Reference
Whey	0.36	0.7	–	0.10[1]	–	[30]
	–	0.25	–	0.03	–	[33]
	0.40	0.25	–	0.03	0.0004	[52]
Butyric acid	0.0113	0.000085[2]	0.17	0.013	–	[14]
	0.0096	12.5	–	0.051	0.0004	[8]
	0.0154	0.33	–	0.083	0.00011	[25]
	0.011	0.032[2,3]	–	0.05	0.0004	[52]
Propionic acid	0.0108	0.0008[2]	0.22	0.023	–	[14]
	0.0033	21.4	–	0.023	0.0004	[8]
	0.013	1.6	–	0.06	0.0004	[25]
						[33]
	–	0.003	–	0.06	–	[52]
	0.005[3]	0.0074[2,3]	–	0.04	0.0004	
Acetic acid	0.0145	0.003[2]	0.33	0.043	0.018	[14]
	0.0142	15.6	–	0.041	0.00013	[8]
	0.0142	5.16	–	0.0567	0.0006	[25]
						[69]
	0.01[4]	0.1[2,4]	1.4[2,4]	–	–	[52]
	0.008[3]	0.1[2]	1.4[2]	0.05	0.0004	
Hydrogen[5]	0.0583	0.0375	–	0.021	0.0004	[8]
	0.0583	0.038	–	0.0283	0.0004	[25]
	–	0.001	–	–	–	[39]
	–	0.0002	–	0.03	–	[33]
	0.058	0.001	–	0.026	0.0004	[52]

[1] Assumption 50% content of C in biomass; [2] free acid; [3] experiment; [4] mean value; [5] dimension for K_S [mol m^{-3}], $Y_{X,S}$ [C-mol mol^{-1}]

Costello et al. [12] used Monod-type kinetics to determine substrate uptake rates, and the growth rates were related to either the product rates or the substrate rates. Hydrogen regulation and inhibition was considered for all acid-forming groups in the manner of Mosey [39] by multiplying the rates of substrate and product formation by hydrogen pressure functions in such a way that the stoichiometry was satisfied. Inhibition by pH was achieved by multiplying the substrate rates by a suitable factor. Product inhibition was modeled with two modified Monod forms on the substrate uptake rates, using the respective acid product concentration for the four groups.

A companion paper [13] tested the model against three different sets of anaerobic reactor data from the literature. The authors found the data from the experiments in the literature to be insufficient for testing the model completely. Particularly lacking was influent alkalinity as well as effluent organic and inorganic carbon, which would have made a carbon balance possible. The following conclusions could be made: hydrogen inhibition at high partial pressures was found to require adjustment of various parameters, the importance of lactic acid as an intermediate was confirmed and well-modeled, it was suggested that other bacterial groups may be required for a complete model, adaptation to low pH was not modeled and may be required, and a 10 day solids residence time described biofilm systems well.

7 Ion Charge Balance

Calculation of the available undissociated organic acid substrates involves considering the pH, which greatly influences anaerobic kinetics.

A recent review of anaerobic modeling concepts [36] has stressed the importance of three buffer systems to determine the pH: carboxylic acid-bicarbonate, acetic acid-acetate if bicarbonate ions are absent, and ammonia-ammonium if carbon dioxide is absent. The Mosey model [39] utilized these to model operation over a wide range of pH with a computer solution that chose the dominating buffer system according to the reactor conditions. The pH sensitivity of the organism growth was modeled by a simple variable factor and the organism death rate pH sensitivity.

Costello et al. [12] used the Mosey model [39] as a basis but also included lactic acid as an important intermediate. Dynamic mass balances for organic acids and inorganic carbon were used. The equilibrium relations described the dissociation of carbon dioxide and all organic acids. A dynamic sodium ion balance was used in conjunction with a charge balance to calculate pH.

In any liquid element an ion charge balance yields

$$\sum (C_{\text{cation}_i} \times \text{charge}_i) = \sum (C_{\text{anion}_j} \times \text{charge}_j) \tag{9}$$

In the pH range of interest (around pH = 7) all strong acids and strong bases are completely dissociated. Weak acids and bases are only partly dissociated.

$$AH \rightleftarrows A^- + H^+ \tag{10}$$

The degree of dissociation is determined by the dissociation constant as

$$K_A = \frac{C_{A^-}H^+}{C_A} \tag{11}$$

where: C_A – concentration of the undissociated acid; C_{A^-} – concentration of the corresponding base (dissociated acid).

Moderately strong acids occurring in anaerobic systems are: acetic: $K_{Ac} = 1.73 \times 10^{-2}$ [mol m^{-3}], propionic: $K_{Pr} = 1.31 \times 10^{-2}$ [mol m^{-3}], butyric: $K_{Ac} = 1.44 \times 10^{-2}$ [mol m^{-3}], carbon dioxide: $K_{Ac} = 3.02 \times 10^{-2}$ [mol m^{-3}], hydrogen sulfide: $K_{H_2S} = 1.26 \times 10^{-4}$ [mol m^{-3}] and ammonium: $K_{NH_4^+} = 5.28 \times 10^{-7}$ [mol m^{-3}]. The only moderately strong base is ammonia: $K_{NH_3} = 1.85 \times 10^{-2}$ [mol m^{-3}]. At pH values below 7 ammonia is practically completely dissociated ($c_{NH_3} \leq c_{NH_4^+}$*0.005). Correspondingly, the buffering effect of ammonia is negligible. The existing cations in anaerobic degradation process are therefore cations (K^+) from strong bases (e.g. Na^+, K^+, NH_{4^+}, Ca^{2+}). In the expected pH range one always has $\Sigma C_{K^+} \gg C_{H^+}$, where ΣC_{K^+} is the total cation concentration. Negative ions are mainly from strong acids (e.g. Cl^-, SO_4^{2-}) and from weak acids (Ac^-, Pr^-, Bu^-, HCO_3^-). The concentration of CO_3^{2-} is always much smaller than that of HCO_3^- (e.g. at pH = 7; $C_{CO_3^{2-}}/C_{HCO_3^-} \approx 4 \times 10^{-4}$). From this one obtains

$$\sum \frac{K_{B_i}}{K_{B_i} + \frac{K_W}{C_{H^+}}} C_{B_{tot,i}} + \sum C_{K^+} = \sum \frac{K_{A_i}}{K_{A_i} + C_{H^+}} C_{A_{tot,i}} + \sum C_{An^-} \tag{12}$$

K_{A_i} – Acid dissociation constant (e.g. K_{Ac}); K_{B_i} – Base dissociation constant (e.g. K_{NH_3}); K_W – Dissociation constant of water; $C_{B_{tot,i}}$ – total concentration of base i, i indicates ammonia; $C_{A_{tot,i}}$ – total concentration of acid i; i indicates CO_2, acetic, propionic, butyric acid and hydrogen sulfide. From Eq. (12) the pH can be estimated for any situation solving a non-linear equation provided the total concentrations of weak acids ($C_{A_{tot,i}}$), weak bases ($C_{B_{tot,i}}$), cations of strong bases (C_{K^+}) and of anions of strong acids (C_{An^-}) are known.

From the above terms, CO_2 is always important. If at 1 bar total pressure 30% of the gas is CO_2, the equilibrium liquid phase concentration will be around 4 [mol m^{-3}]. At pH below 7.25, less than 1% of ammonia nitrogen exists in the free base form. If the total concentration of ammonia is rather low (< 1.5 [mol m^{-3}]) the corresponding terms can be neglected. The H_2S/HS^- system has its highest buffering capacity at pH 6.9. If the pH is below 6, this buffer system is not important. The experiments described by Ryhiner [52] were typically around pH ≈ 6, which reduces the importance of this system for buffering. The concentration of sulfur in whey waste water is quite low. The measured gas concentrations were usually below 2%. The corresponding H_2S concentration in the liquid phase was therefore always < 0.6 [mol m^{-3}].

Neutralization reaction is a very fast reaction reaching equilibrium almost instantaneously. During pH control strong base is added. The addition of strong alkali causes an increase in ΣC_{K^+} which, by solving Eq. (12), results in a decrease of C_{H^+}.

It is possible to use only the difference between cations and anions

$$C_Z = \sum C_{K^+} - \sum C_{An^-} \tag{13}$$

For the numerical solution it is useful to rearrange Eq. (12) and use Eq. (13) in the form,

$$\delta = \sum \frac{K_i}{K_i + C_{H^+}} C_{tot,i} - \sum \frac{K_{B_i}}{K_{B_i} + \dfrac{K_W}{C_{H^+}}} C_{B_{tot,i}} - C_Z \tag{14}$$

C_{H^+} is then varied iteratively until δ approaches sufficiently closely a value of 0. This numerical solution is not always trivial using conventional methods for non-linear algebraic equations (e.g. Newton–Raphson, Regula falsi as incorporated in the simulation language ACSL). Bellgardt [7] used a very robust algorithm which has been found to converge in any case. The iteration shown in Fig. 2 was started at $C_{H^+} = 10^{-14}$ kmol m^{-3}. C_{H^+} was then increased by a factor f_n (usually 10) until $\delta < 0$. This corresponds to moving along one single curve in Fig. 2 until the line intersects with $\delta = 0$. Then C_{H^+} was decreased by f_n and $f_{n+1} = \sqrt{f_n}$. Now, the previous procedure was repeated again until $\delta < 0$.

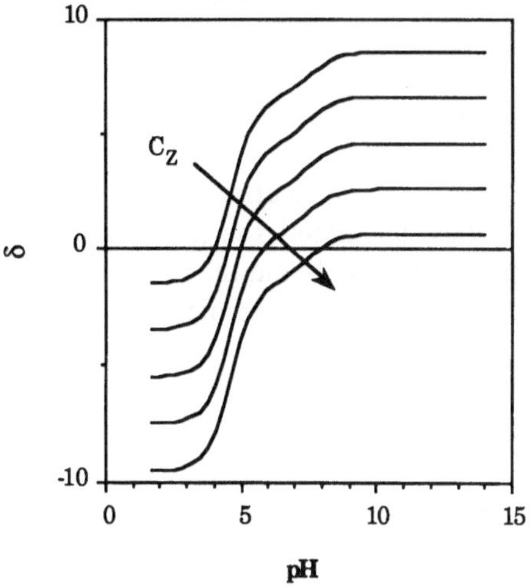

Fig. 2. Estimation of pH for different cation concentrations C_Z using Eq. (22) $C_{Ac} = 4$ mol m^{-3}; $C_{CO_2} = 4$ mol m^{-3}; $C_{H_2S} = 0.6$ mol m^{-3}; $C_{NH_3} = 4$ mol m^{-3}; $C_Z = 0, 2, 4, 6, 8$ mol m^{-3}. The *arrow* indicates increasing cation concentration

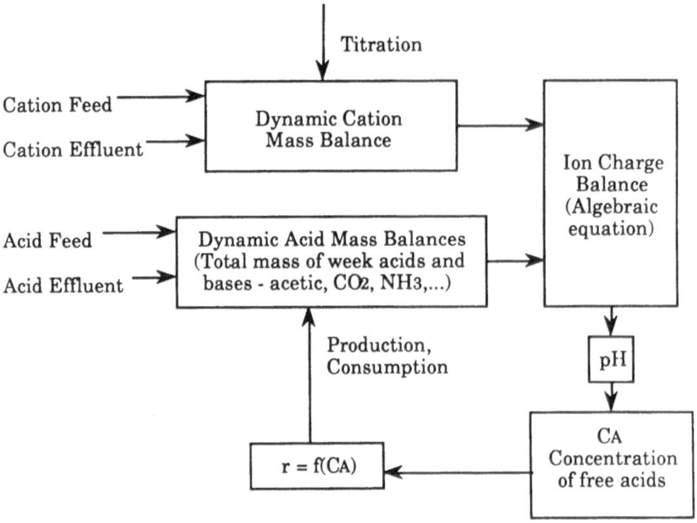

Fig. 3. Schematic of the ion charge balance iteration within a dynamic simulation. C_A – free acid concentration

The iteration was stopped when reaching an accuracy criterion, usually $\left|\frac{\delta}{C_{H^+}}\right| \leq 0.01$. Convergence is always guaranteed. If initially pH but not C_Z is known, C_Z can be estimated from Eq. (14) with $\delta = 0$.

In addition to the mass balance equations for total weak acids (acetic, propionic, butyric, carbonic, ...) and weak bases (ammonium), a balance for cations of strong bases (K^+, Na^+, ...) and anions of strong acids (Cl^-, SO_4^{2-}, ...) is necessary as follows:

$$V_L \frac{dC_Z}{dt} = F_L\,(C_{Z_0} - C_Z) + F_{titr}\,C_{Z_{titr}} \tag{15}$$

Figure 3 gives an information flow diagram of how the solution of the dynamic mass balances depends on the reaction kinetics. The kinetics in turn depend on the undissociated acids, which are given by the total acids and the pH, according to the equation. The charge balance represents an algebraic loop in the otherwise sequential dynamic integration.

8 Modeling Biofilm Kinetics

In those reaction systems where biofilms or flocs are formed (UASB, fluidized bed, fixed bed) diffusion and direct inter-species transfer of intermediates (e.g. hydrogen) may be important. The detailed biofilm structure and formation is very complex which can easily be seen in electron microscopic pictures. It can be assumed that the biofilm structure is also dynamic but with longer time constants than the hydraulic retention time. Quantitative data on biofilm structure and inter-species hydrogen transfer are practically not existing [36]. It is also difficult to determine a representative diffusion coefficient. Examples of biofilm modeling in the literature are: Denac et al. [17]; Bryers [8]; Wang et al. [65].

9 Gas Phase Modeling

The transport of gaseous components (methane, hydrogen and carbon dioxide) from the liquid phase to the gas phase can be modeled as a well-mixed gas phase with mass transfer driven by the difference between solubility and liquid phase concentration [52]. The flow rate leaving the reactor was calculated from the bubble velocity and gas volume fraction considerations. Thus,

Accumulation = Transport − Gas flow

$$\frac{dn_i}{dt} = V_L k_L a(x_i H_i p - c_{Li}) - \frac{x_i u_B A \varepsilon_g p}{RT} \tag{16}$$

In addition the time delay of the gaseous components due to the head space was modeled with a total gas balance. The gas liquid mass transfer coefficient, $k_L a$, was modeled as a function of fractional gas hold-up, which was a function of the gas production rate. Mather [33] used a similar concept to model the gas phase of an anaerobic reactor.

10 Comparison of Simulations with Experiment

It is very difficult to set up a model which qualitatively and quantitatively agrees with experimental data. Therefore, comparisons of experimental data with simulation results are rather scarce.

The model of Smith and McCarty [57] predicted the cyclic methane production that occurred after shock loading, due to the inhibition of hydrogen on the propionate conversion to acetate and due to the biomass population dynamics in the continuous tank reactor with suspended culture. This work has shown that the present knowledge of the detailed kinetics is still not adequate to establish a complete model. Tschui [62] set up a complex model of anaerobic sludge stabilization and compared simulated results with substrate pulse experiments. The qualitative agreement of all parameters was reasonable.

The model of Ryhiner [52] was used to compare simulated results with experimental data from whey treatment. Steady state and dynamic comparison was made as given below.

10.1 Steady State Comparison of Experimental and Simulated Carbon Balances

From steady state simulations and experiments, carbon fluxes were determined, and carbon balances were estimated. The amount of carbon in the feed was taken as reference (=100%). A comparison of these values for the single stage system gave good agreement between experiment and simulation. The deviations for the total carbon balance were below 5%. Also other values (methane and CO_2 in the gas phase, dissolved CO_2, carboxylic acids) agreed very well. The total carbon balances for the two stage system also agreed well but showed larger deviations for individual compounds.

10.2 Comparison of Dynamic Experimental and Simulation Results

Dynamic responses to step changes in feed concentration (+42%) were investigated experimentally and by simulation. Examples from the single stage reactor are shown for pH response in Figs. 4 and 5 and the response in the gas composition in Figs. 6 and 7. The response in dissolved hydrogen concentration (Fig. 8) was only available from simulation.

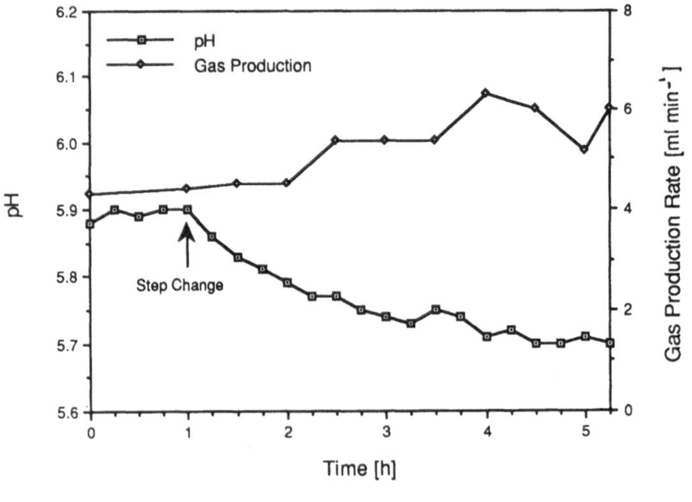

Fig. 4. Response of pH and gas production rate on concentration step change (1.4 kg m^{-3} → 2.0 kg m^{-3} whey powder) in single stage reactor (experiment)

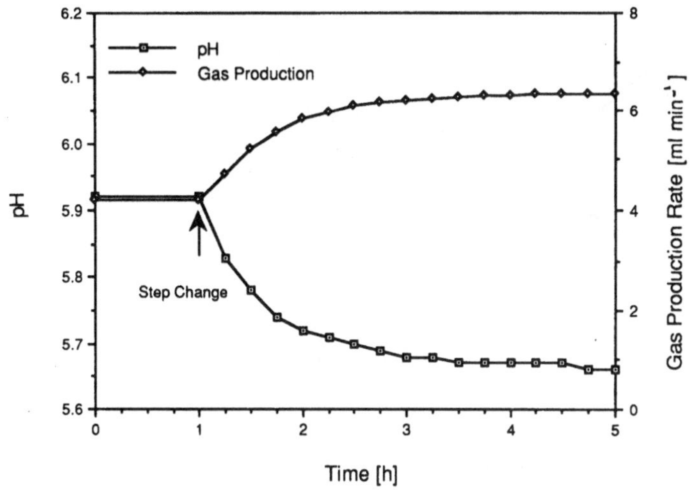

Fig. 5. Response of pH and gas production rate on feed concentration step change (1.4 kg m^{-3} → 2.0 kg m^{-3} whey powder) in single stage reactor (simulation)

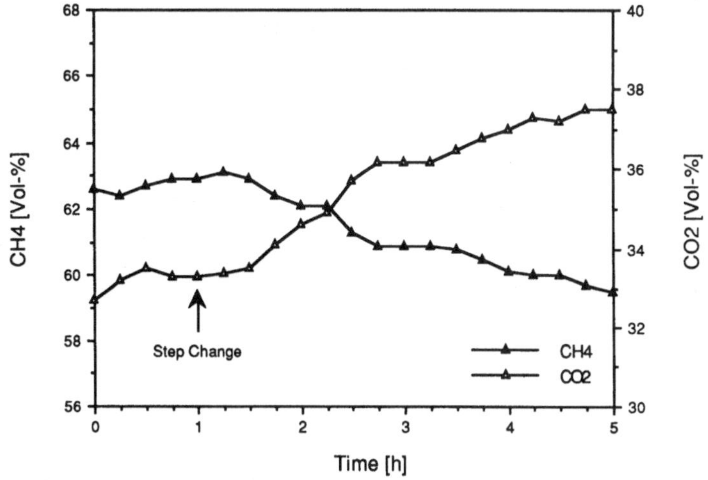

Fig. 6. Response of gas composition on feed concentration step change (1.4 kg m^{-3} → 2.0 kg m^{-3} whey powder) in single stage reactor (experiment)

11 On-line Measurement and Observation Methods

The off-line and on-line measurable variables for anaerobic methane-producing processes have been extensively reviewed [60]. The variables considered for on-line application were the following: pH, organic acids, alkalinity, redox, poten-

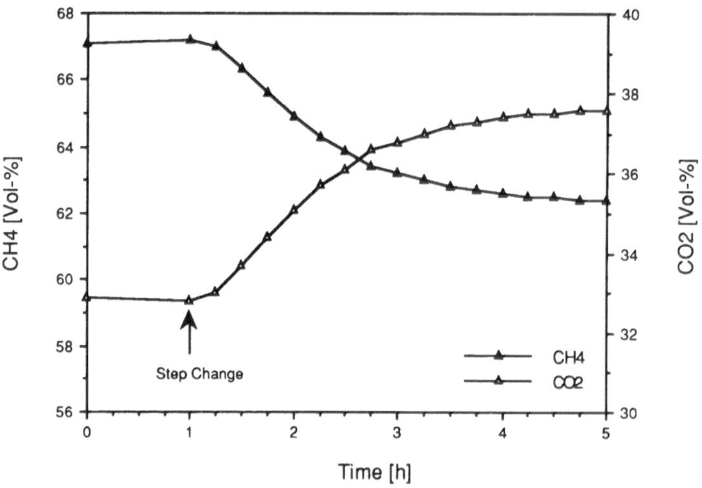

Fig. 7. Response of gas composition to a feed concentration step change increase (1.4 kg m^{-3} → 2.0 kg m^{-3} whey powder) in single stage reactor (simulated)

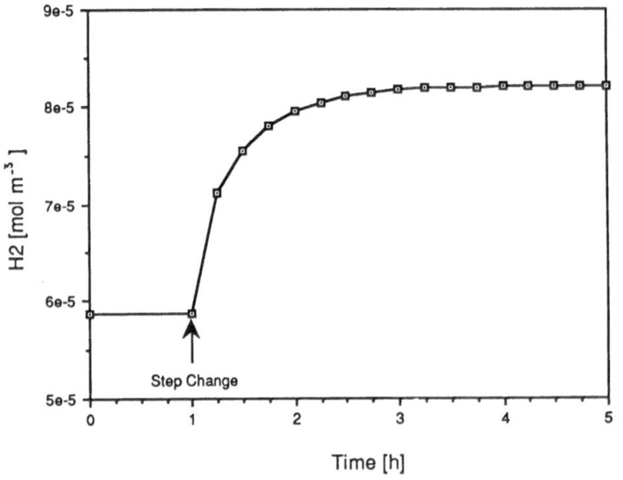

Fig. 8. Response of dissolve hydrogen to a feed concentration step change increase (1.4 kg m^{-3} → 2.0 kg m^{-3} whey powder) in single stage reactor (simulated)

tial,sulfide ion activity, gas production rate, gas composition, hydrogen gas, and carbon monoxide gas. Bicarbonate alkalinity is determined by acid titration to defined pH between ($4.0 \leq pH \leq 5.75$), as discussed in detail by Powell and Archer [44] and Di Pinto et al. [43]. Switzenbaum et al. [60] felt that carbon monoxide and hydrogen gas responded sufficiently fast to overloads to make them useful for control. A later report by the same authors [27] described their experimental work on carbon dioxide and hydrogen responses and concluded that CO gave information on the acetate consumption, while H_2 was involved mainly in the CO_2 reduction step. Pauss et al. [40] measured continuously the dissolved hydrogen in the liquid phase using a commercial hydrogen/air fuel cell-based membrane probe. This probe was highly sensitive to hydrogen but exhibited interference with H_2S in high sulfur containing waste streams. It was further shown that dissolved hydrogen concentration deviated from gas phase equilibrium by a factor of about 50. The gas phase hydrogen sensor used by Tschui [62] required absorption of H_2S to avoid poisoning but was useful to follow substrate pulses in mesophilic anaerobic sludge stabilization.

Denac et al. [17] utilized the NaOH consumption rate of the pH control system to devise a control scheme which manipulated the influencing feed rate to the reactor. Since the algorithm required turning off the feed during overloads, equalization tank capacity would be required. In addition this principle was used to maintain a constant range of effluent organic acids concentration.

Schürbüscher and Wandrey [55] listed on-line measurement methods for anaerobic digestion including experimental problems and benefits of each method. They also showed the application of a Kalman filter to estimate non-measurable variables.

12 Control of Anaerobic Processes

Control of anaerobic processes may be necessary because of possible detrimental disturbances of the process by overloading or toxification. Applied control algorithms were mostly of simple conventional PID type and there are only few experimental results in the literature.

Several variables were manipulated by the controllers applied. Most controllers acted on the feed flow. This, however, requires a buffer tank to allow intermediate storage of the wastewater. Another control strategy uses addition of a base (e.g. $CaCO_3$) to keep pH at a desired value. This creates significant operational costs. Another suggested method of pH control uses separation of CO_2 from the biogas and recycling of methane to strip CO_2 from the reactor. No experimental results, however, have been reported. CO_2 separation also may be quite costly. Simple recycling of the product biogas without CO_2 removal gives only a very limited possibility of increased rate of stripping CO_2 by increasing k_La. A forth strategy uses controlled sludge recycling. This may be used in contact processes, where surplus sludge is stored in a special tank for later reuse for control.

Various variables were used as measured process output variables. In principle any variable giving information on intermediates like acids, hydrogen, CO_2, etc. can be used as measured variables. pH measurement is quite simple, and it can be kept at a desired set point value by direct manipulation of one of the above mentioned variables. Bicarbonate alkalinity is determined by titration to a pH value between pK of organic acids and carbonic acid. It was shown to give a fast and sensitive response which can be used for control of anaerobic processes [50].

Partial pressure of CO_2 (p_{CO_2}) in the off-gas was reported to be a good stability indicator but a bad control variable [50]. The same was found for p_{CH_4} [52]. Hydrogen partial pressure in the off-gas requires very sensitive sensors but provides fast information on the process condition [71]. It was claimed that there is no long-term response [38]. Dissolved hydrogen concentration was measured by mass spectrometer and made control of anaerobic processes possible [66, 68, 31]. Volatile fatty acid accumulation is a clear indication of overloading or toxification and allows control of an anaerobic bioreactor. The measurements are, however, expensive and relatively complex involving on-line GC measurement of the liquid phase [46, 52, 72]. Other controllers use model based estimation of controlled variables. This includes the application of a Kalman-filter as described by Schürbüscher and Wandrey [55]. Table 4 lists recent simple-type control studies. Of these 15 studies, 9 were experimental (4 pilot or industrial scale, 5 laboratory scale) and 6 involved simulations. One study involved both simulations and lab-scale experiments.

12.1 Simple Controller Design using Simulations of a Complex Process Model

As experiments with anaerobic systems are very time consuming, it is desirable to design controllers using a complex process model. An example of such a study [52, 53] will be described in more detail below. The general procedure is sketched in Fig. 9. The control was implemented as PID-type feed back control using a discrete digital control algorithm in the position form. The sampling time was 15 min which allowed for complete analysis of gas (CO_2 and CH_4) and liquid phases (acetic, propionic and butyric acid). The simulation model permitted the determination of the suitable sampling time and controller settings by the Ziegler–Nichols method.

Since setting the controller parameters required additionally a certain amount of trial and error, the simulation model was valuable in determining an approximate setting. The integral and differential settings were identical in experiment and simulation, whereas the proportional constants needed re-adjustment.

12.2 Proportional-Integral-Differential Control of pH

The control of pH by manipulating the feed rate using PID digital control was investigated to counteract a step change in the feed concentration in single and two stage fluidized bed reactors. For the single stage system the feed rate was increased by a factor of two and for the two stage system by a factor of three. The differential part was found to be ineffectual. Comparisons of the experimental and simulated PI control responses for step changes in the flow rate are given in Figs. 10 and 11 for the single stage. Here the controller settings were essentially

Table 4. Examples of simple control studies in the literature

Control variable	Control action	Controller type	Advantages (+) disadvantages (−)	System	Ref.
pH	Feed flow	P	(−) Holding tank	Anaerobic contact	[10]
	Recycle of sludge from a 2nd stage	P	(−) Restriction of control action	process Simulations	[10]
	CO_2 scrubbing and gas recycle				[1]
	Base addition	On-/Off			[32] [33]
	Base addition	On-/Off	(−) robust	Primary sediment-	[14]
	Feed flow	PI	(−) efficient	ation (solids) and digester Simulations	[58]
	Feed flow	On-/Off		Anaerobic filter Industrial scale	[45]
	Feed flow	PI	(+) simple, fast (−) depending on buffer capacity	One- and two-stage anaerobic fluidized bed reactors Lab scale and simulations	[52]
pH or Bicarbonate alkalinity (BA)	Base addition	On-/Off	(+) BA faster than pH	Anaerobic contact process	[50] [51]
pH and p_{CO_2}	Base addition	On-/Off	(+) fast response of p_{CO_2}	Simulations	
p_{CO_2}	Base addition	On-/Off	(−) poor control variable	Anaerobic contact process Simulations	[51]
p_{CH_4}	Feed flow	PI	(−) poor control variable	Anaerobic fluidized beds Lab scale	[52]
Rate of methane production	Recycle of sludge from a 2nd stage	On-/Off	(+) at presence of toxics	Anaerobic contact process Simulations	[1]

Table 4. (continued)

Control variable	Control action	Controller type	Advantages (+) disadvantages (−)	System	Ref.
H_2 (gas) and pH	(step response)		(+) fast response of hydrogen (+) long-term response of pH	Anaerobic filter Lab scale	[38]
H_2 (gas)	(step response)		(+) simplicity of measurement (−) transfer liquid-gas	Anaerobic contact digester Pilot scale	[3]
H_2 (liquid)	Feed flow	On-/Off	(+) fast sensitive (−) mass spectrometer required	Anaerobic digester Lab scale	[66]
	Feed flow	PI		Simulations	[52]
H_2 (gas)	(step response)		(−) not sensitive	Anaerobic filter Industrial scale	[71]
Volatile acid to alkalinity ratio	(step response)		(+) simple measurement (titration) (+) no calibration	Anaerobic digester Lab scale	[47]
Bicarbonate Alkalinity	Alkali addition	P	(+) simple measurement (titration)	Anaerobic digester Lab scale	[43]
Volatile acid concentration	Feed flow	Adaptive controller	(−) knowledge of influent concentration and stoichiometry	Anaerobic digester Pilot scale	[18] [46]
	Feed flow	Generic model control	(−) knowledge of influent concentration and stoichiometry	Anaerobic digester Simulations	[72]
	Feed flow	PID	(+) fast (−) on-line measurement of VFA	One- and two-stage anaerobic fluidized bed reactors Lab scale and simulations	[52]

identical. For the two stage system control the controller gain in the experiment was set at 40% of the value found suitable by simulation. In spite of this, some instability was observed, possibly due to the higher buffering in the experiment (twice the amount of dissolved CO_2).

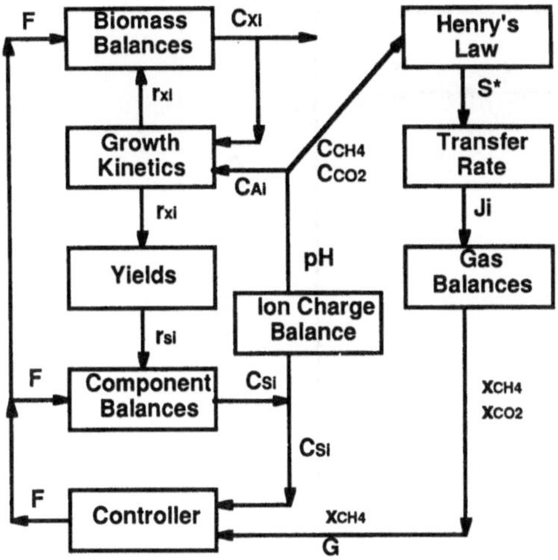

Fig. 9. Calculation flow diagram for the simulation of on-line control of the anaerobic process

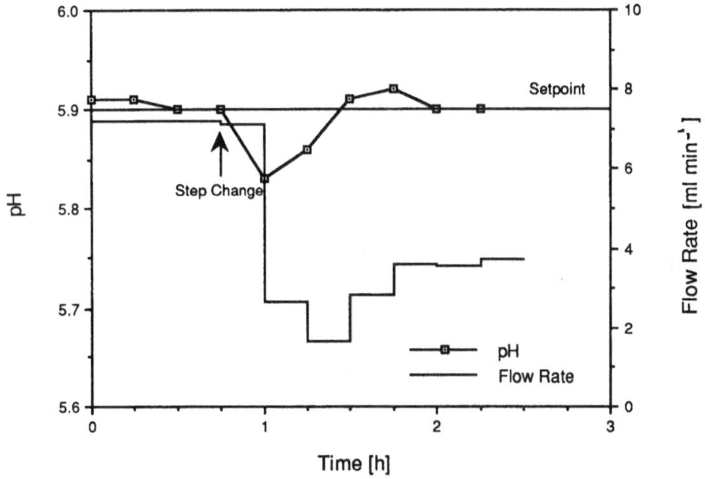

Fig. 10. PI-Control of pH after feed concentration step change ($1.4\ kg\ m^{-3} \rightarrow 2.8\ kg\ m^{-3}$ whey powder), single stage reactor (experiment). ($K_r = 0.7\ m^6\ mol^{-1}\ h^{-1}$, $\tau_I = 0.2\ h$)

12.3 PID Control of Organic Acid Concentrations

The organic acid response was rapid (here with 15 min sampling time) but requires rather expensive analytical methods. Shown here (Figs. 12 and 13) are simulated and experimental results from PID control based on acetic acid measurement; the controller constants were similar but not identical. The gain

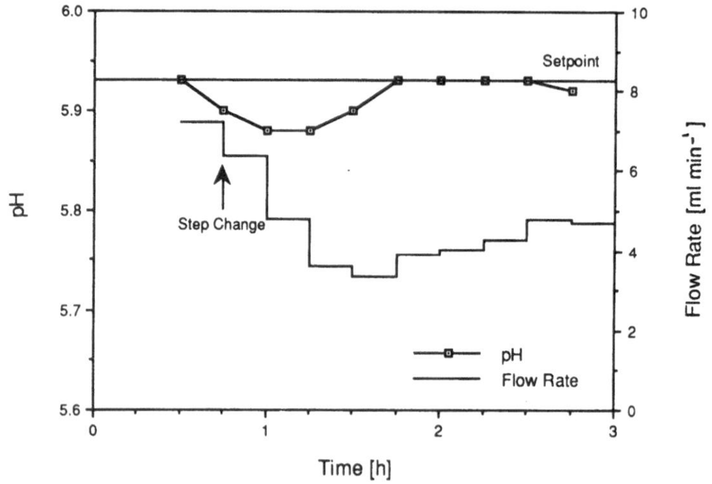

Fig. 11. PI-Control of pH after feed concentration step increase (1.4 kg $m^{-3} \rightarrow$ 2.8 kg m^{-3} whey powder), single stage reactor, simulation. (K_r = 0.6 m^6 mol^{-1} h^{-1}, τ_I = 0.2 h, τ_D = 0.05 h)

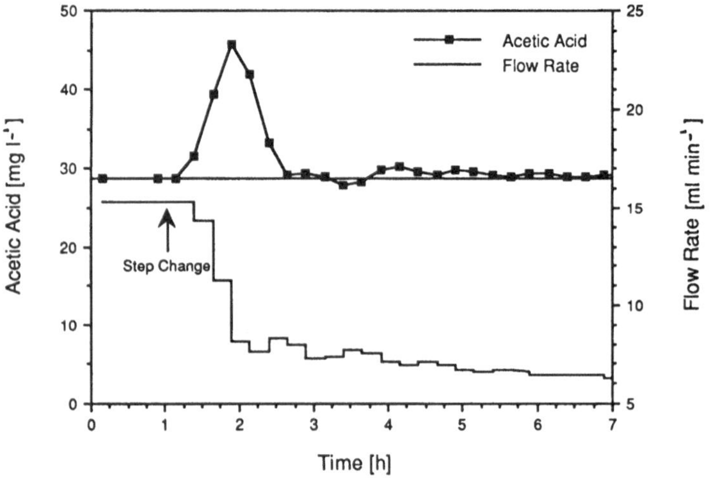

Fig. 12. Acetic acid set-point, PID-control of the two stage system, simulated. ($K_r = 3 \times 10^{-4}$, τ_I = 0.25, τ_D = 0.2). Step change concentration increase (1.4 kg $m^{-3} \rightarrow$ 4.2 kg m^{-3} whey powder)

was set somewhat high in the experiment and caused oscillations. Similar results were obtained with propionic acid control.

12.4 Control of Dissolved Hydrogen Concentration

Hydrogen was not measured, but a simulation of hydrogen control with a PI controller demonstrated that this was a sensitive control variable (Fig. 14).

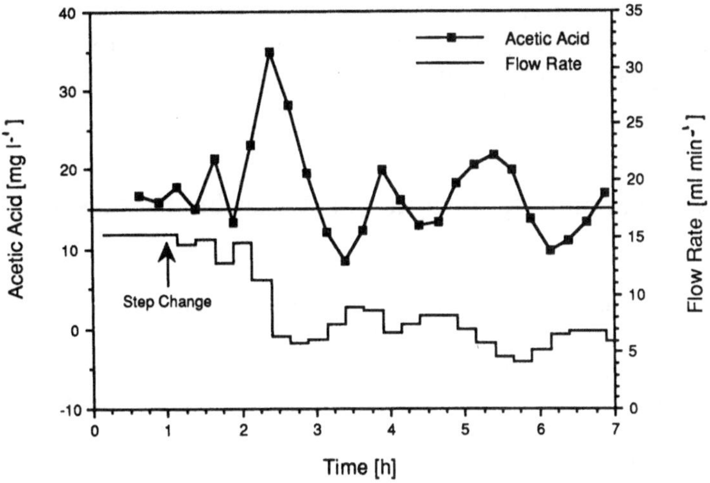

Fig. 13. Acetic acid set-point, PID-control of the two stage system, experimental. ($K_r = 3 \times 10^{-4}$, $\tau_I = 0.3$, $\tau_D = 0.05$). Step change concentration increase ($1.4 \text{ kg m}^{-3} \rightarrow 4.2 \text{ kg m}^{-3}$ whey powder)

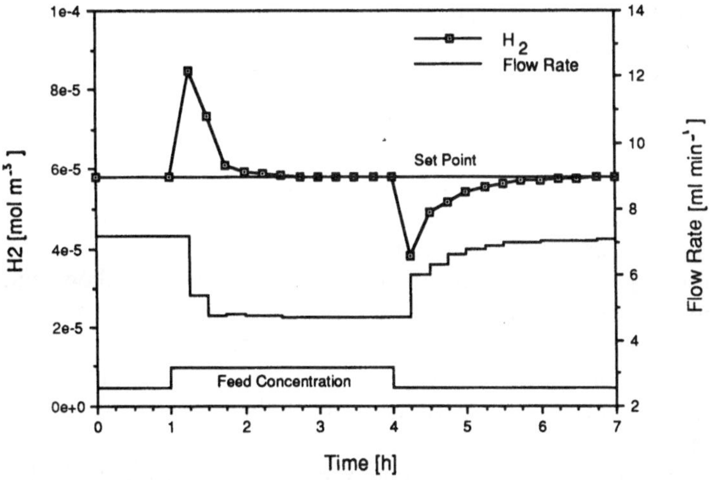

Fig. 14. Simulated response of the PI hydrogen controller to a square wave in feed concentration for the single stage system. ($K_r = 2.5 \text{ m}^6 \text{ mol}^{-1} \text{ h}^{-1}$, $\tau_I = 0.2$ h). Step change concentration increase ($1.4 \text{ kg m}^{-3} \rightarrow 2.8 \text{ kg m}^{-3}$ whey powder)

12.5 Control Based on the Methane Content of the Gas

For high rate anaerobic treatment, the liquid phase flow rate greatly influences the amount of carbon dioxide in the gas phase. Thus the difference in solubilities of methane and carbon dioxide makes this type of feedback control difficult to implement. In principle a computer control system could correct for the dissolved carbon dioxide in the effluent. Basing the control on the mass flow of methane in the gas phase could be more successful.

13 Adaptive Control and Optimization of Anaerobic Processes

Optimal operation of anaerobic process should allow minimization of fixed, variable and operating costs according to a defined performance index. Basically two types of optimization method may be used. Steady-state methods estimate the steady state gain and use a gradient method with steepest descent to approach the variable optimum [49, 52, 53]. A general disadvantage of this method is that it is relatively slow, but uses simple linear models as described later for the case of whey treatment. Dynamic methods [42] or model-based methods [24, 56] allow a faster response at the expense of a more complex model setup. Only steady state methods have been applied for anaerobic digestion processes as far as has been reported.

On-line measurements of the variable to be controlled permit taking control action with simple conventional feed-back methods. However, the characteristics of anaerobic treatment of wastewater continually change, and such simple methods are therefore not always successful. Set-point control is also usually not sufficient when it is of interest to maintain the process at an optimum which is a function of two or more process variables. In such cases a model for the process is necessary if the controller is to adapt to changing conditions. In adaptive optimal control, a model is used for a changing process in order to seek a maximum in a performance index function.

A recent book [6] reviews the principles of adaptive control methods and their application to biological reactors. The literature contains relatively few applications of optimization methods for anaerobic wastewater treatment systems, most systems are simulated, and experimental work is lacking (Table 5). These systems are of particular interest because of the potential for widely varying feed composition and biomass adaptation.

A four organism model Mosey-type model was used by Dochain et al. [19] to develop an adaptive control algorithm for the control of hydrogen concentration, which was tested by simulation. In this control scheme the hydrogen level was to be controlled by measuring the inflow glucose, the outflow hydrogen gas and the liquid phase hydrogen concentration on-line and by manipulating the inflow rate. A quasi-steady state assumption regarding the glucose concentration allowed relating the accumulation rate for hydrogen to the glucose inflow rate and the hydrogen outflow gas rate. A further assumption neglected the propionate reaction path. It was possible to prove mathematically the stability of the proposed control system without knowing the kinetics or yields. For the simulation both hydrogen and acetate were assumed to have inhibiting effects

Table 5. On-line optimization of anaerobic processes

Control action	Performance index	Optimization	System	Ref.
Recycle ratio	Cost function: fixed costs (HRT) and variable costs (organics and substrate in the effluent)	Simplex search Off-line	Two-stage anaerobic digester with settler Simulations	[23]
Feed flow	Deviation of pH and feed flow rate	Fibonacci search On-line	Anaerobic digester Simulations	[10]
Base addition (feed-forward/ feed-backward control)	same plus base flow			
Feed flow (residence time)	Biogas production (space-time yield)	Closed-loop On-line	Chemostat Pilot scale	[64]
Feed flow	Productivity	Adaptive on-line optimization, Steepest descent	Bakers' yeast fermentation Lab scale	[49]
Feed flow (residence time)	Biogas production (space-time yield)	Closed-loop On-Line	Chemostat Pilot scale	[64]
Feed flow	Gas production rate	Empirical questing routine	Stirred tank laboratory reactors	[33]
Feed flow	Methane production minus VFA in the effluent	Adaptive on-line optimization, Steepest descent	One and two-stage anaerobic fluidized bed reactors Lab scale and simulations	[52]

on the methanogenic steps and the neglected propionate reaction was included. Without control the hydrogen consuming methanogens are washed out of the reactor. Numerous simulations tested the controller under a range of conditions, including a comparison with simple PI control. This approach appears to be a promising way of including some of the model complexities into an adaptive controller design.

Detailed mechanistic models for adaptive control are generally too complex, and would require simplification before implementation. However they have the advantage of containing most information and may be necessary if not all variables can be measured. When all the important process variables can be measured, then empirical models of a simple, linear form can be used to represent the relations between manipulated inputs and measured outputs [49].

$$y_1(t) + a_{1,1}\,y_1(t-1) + a_{1,2}\,y_1(t-2) + \ldots\ldots + a_{1,n}\,y_1(t-n)$$
$$= b_{1,1}\,u(t-1-d_1) + c_1 \tag{17}$$

$$y_2(t) + a_{2,1}\,y_2(t-1) + a_{2,2}\,y_2(t-2) + \ldots\ldots + a_{2,n}\,y_2(t-n)$$
$$= b_{2,1}\,u(t-1-d_2) + c_2 \tag{18}$$

where d_1 and d_2 are integer indices which refer to the number of sampling times that the output is delayed, due for example to the residence time of the process or to the measurement delay. The nomenclature $(t-n)$ refers to the sample taken n sample times earlier. The sample time is determined by the process dynamics and the analytical instruments. The input u may be, for example, feed flow rate; the outputs y_1 and y_2 could be measured concentrations or uptake rates.

Such empirical models have the advantage that very little process understanding is necessary, other than knowing what to measure and what to manipulate. The theory behind the method allows the process steady state to be estimated from the dynamic data, making it unnecessary to wait for the steady state. The optimal inputs are found from this information, which is an important advantage for slow biological processes.

The elements of an adaptive algorithm are summarized in Fig. 15.

A complex mechanistic simulation model as described earlier was used to find the maximum of the performance index (PI), which was a compromise between high methane production and low effluent acids concentration [52].

The feed flow rate was adjusted to maximize a desired performance index, taken as a compromise between high methane production and low effluent

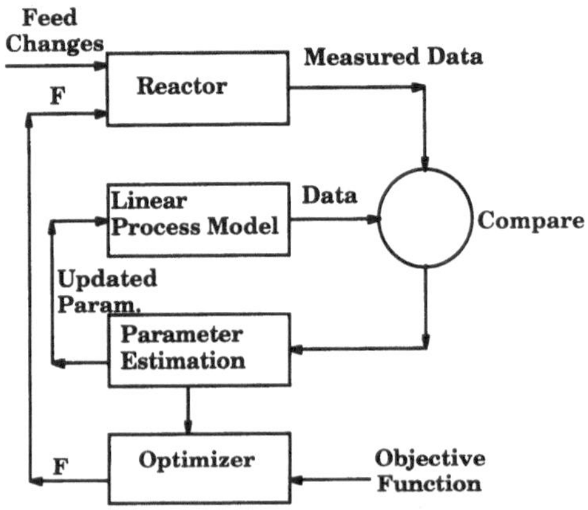

Fig. 15. Continual parameter updating allows the use of a simple model. The optimizer calculates the flow rate required to reach the maximum

concentration. In this application the aim was to maximize the rate of anaerobic degradation without allowing the organic acid concentrations to become excessively high. Thus the performance index, PX, was formulated as follows:

$$\begin{bmatrix}\text{Objective}\\ \text{function}\end{bmatrix} = \begin{bmatrix}\text{Methane}\\ \text{production}\\ \text{rate}\end{bmatrix} - \text{Constant} \times \begin{bmatrix}\text{Total}\\ \text{organic acids}\\ \text{concentration}\end{bmatrix}$$

$$PI = y_1 - \beta \times y_2 \tag{19}$$

where $y_1 = r_{CH_4}$ and $y_2 = \Sigma c_{Ai}$. The shape of the performance index was found by the use of the differential equation simulation model of the process. From this the influence of feed rate on the performance index at steady state could be determined. This function, as shown in Fig. 16 is dependent on feed concentration and feed rate.

Simulation results of the reactor with the estimation (Eqs. 17 and 18) indicated that the simplest model was sufficient for obtaining good estimates of y_1 and y_2.

$$y_1(t) = b_{1,1}\, u(t - 1 - d_1) + c_1 \tag{17}$$

$$y_2(t) = b_{2,1}\, u(t - 1 - d_2) + c_2 \tag{18}$$

This was therefore used in the experiments. This simple model considers only the most recent output data directly, and although much of the dynamic information was not considered by this simplification, this was found to be of no disadvantage. Influence of older data was included by variable forgetting factor for parameter values.

Simple input-output models were used for the adaptive optimization of one and two stage anaerobic fluidized bed reactors.

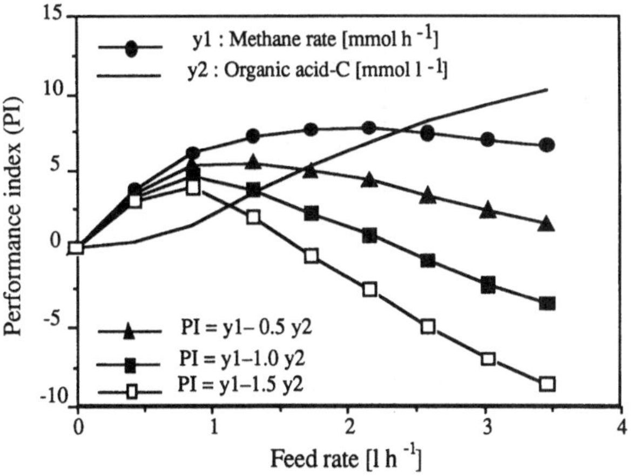

Fig. 16. The performance index depends on the penalty function constant for the organic acid concentration, here for a single stage system, simulated (Feed concentration = 42 C-mol m^{-3})

The strategy for control was as follows:

1. Sampled data on organic acid and methane concentrations (GC) and gas flows were used to manipulate the feed flow rate.
2. Control by adaptive on-line optimization was implemented using a linear, single input (feed flow rate) – double output (methane rate and total organic acids concentration) model. Thus as given in Eqs. (17) and (18), the feed flow rate (u) was manipulated and caused changes in both the methane rate (y_1) and total organic acids concentration (y_2). The parameters of the model b_i and c_i were continually updated to give a fit to the current time period. The model was then used to calculate which feed flow changes would move the process toward the desired optimum. If the process changed, say, due to a change in substrate feed composition, then the outputs would move away from the optimum. This would be sensed, incorporated into the model parameters, and the feed flow rate would be changed to move towards the new optimum.

The values for the controller parameters were found by simulation using the mechanistic model; they compared favorably with the values found from experiment, with the differences probably due to influence of measurement noise. Simulated results for the single stage reactor are given in Figs. 17 and 18. In the simulation the optimizer quickly responded to the feed concentration increase disturbance keeping the performance index (PI) nearly constant. Feed concentration decrease did not allow keeping the performance index constant. Instead of that the flow rate was kept constant and only increased after

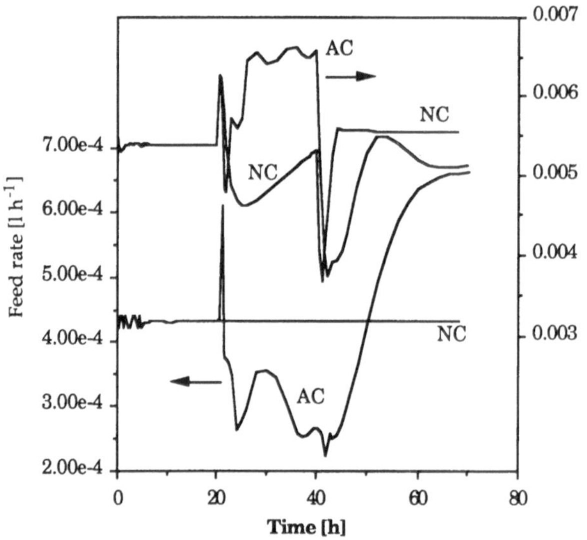

Fig. 17. Simulated optimizer response of the single stage process to a step change (42 to 63 C-mol m^{-3}) in the feed concentration. AC – adaptive control, NC – no control. ($\alpha = 10^{-6}$, $\beta = 1.0$, $\Sigma_{0,1} = 10^{-7}$, $\Sigma_{0,2} = 10^{-7}$, $\lambda_{min} = 0.1$)

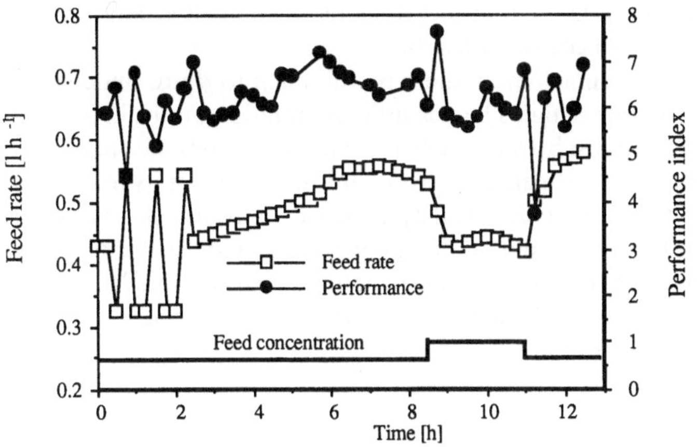

Fig. 18. Controlled response of the single stage process to a step change (42 to 63 C-mol m^{-3}) in the feed concentration (experiment). ($\alpha = 2 \times 10^{-6}$, $\beta = 1.0$, $\Sigma_{0,1} = 10^{-6}$, $\Sigma_{0,2} = 10^{-6}$, $\lambda_{min} = 0.1$)

considerable delay. The results obtained using the adaptive controller are compared to uncontrolled system response. A feed concentration increase by 75% (42 → 73.5 C-mol m^{-3}) was handled much better by the adaptive controller. Comparison of response to concentration decrease (time = 40 h) is difficult from Fig. 17 since the starting points were different. In the experiment (Fig. 18) the controller reacted more quickly in both cases. The flow rate decrease by the controller corresponded nearly to keeping constant the substrate loading rate of the reactor. The experimental results for the two stage reactor were similar to the simulated results but had a significant time delay and correspondingly larger overshooting.

It is quite remarkable that the detailed mechanistic simulation model was so useful for designing and testing the adaptive controller. The controlled process could be simulated for all controller-types, allowing the controller parameters to be found. The performance index as defined was satisfactory for control and the algorithm from the literature [49] was usable.

14 Concluding Remarks

Recent developments in modeling, simulation and control of anaerobic processes have been reviewed. Despite the fact that anaerobic processes are very complex multi-organism, multi-substrate processes, it seems to be possible to describe the essential features of anaerobic degradation process dynamics. A relatively complex model has been successfully used to describe dynamic responses to changing feed conditions and to design simple feedback and advanced optimizing controllers. Controller design using models is important in anaerobic processes partly because of lengthy experimentation time.

Although much has been achieved, there is still much to do to understand anaerobic processes in more detail. There is little information on the kinetics of degradation of less common substrates such as aromatic compounds frequently occurring in industrial waste streams. Microorganism interaction is only rudimentarily understood and therefore crudely modeled. It is still not yet clear which method is best for controlling the stable operation of anaerobic reactors. It may be that the method may depend on the individual case. Simple, reliable and robust measurements are required, and they will need more complex control methods to be successful.

15 References

1. Andrews JF (1978) The development of a dynamic model and control strategies for the anaerobic digestion process. In: Mathematical models in water pollution control. J Wiley, Chichester, p 281
2. Archer DB (1983) Enzyme Microb Technol 5: 570
3. Archer DB, Hilton MG, Adams P, Wiecko H (1986) Biotechnol Lett 8: 570
4. Atkinson B, Mavituna F (1983) Biochemical engineering and biotechnology handbook. Nature Press, New York
5. Barnes D, Fitzgerald PA (1987) Anaerobic wastewater treatment processes. In: Forster CF, Wase DAJ (eds) Environmental biotechnology. Ellis Horwood, Chichester, p 57
6. Bastin G, Dochain D (1990) On-line estimation and adaptive control of bioreactors. Elsevier, Amsterdam
7. Bellgardt KH (1991) personal communication
8. Bryers JD (1985) Biotechnol Bioeng 27: 638
9. Colberg PJ (1988) Anaerobic microbial degradation of cellulose, lignin, oligolignols, and monoaromatic lignin derivatives. In: Zehnder AJB (ed) Biology of anaerobic microorganisms. Wiley, New York, p 333
10. Collins AS, Gilliland BE (1974) J Environ Eng Div 100: 487
11. Conrad R, Schink B, Phelps TJ (1986) FEMS Microbiology Ecology 38: 353
12. Costello DJ, Greenfield PF, Lee PL (1991a) Water Res 25: 847
13. Costello DJ, Greenfield PF, Lee PL (1991b) Water Res 25: 859
14. Dalla Torre A, Stephanopoulos G (1986) Biotechnol Bioeng 28: 1106
15. Denac M, Dunn IJ (1988a) Biotechnol Bioeng 32: 159
16. Denac M, Miguel A, Dunn IJ (1988b) Biotechnol Bioeng 31: 1841
17. Denac M, Lee PL, Newell RB, Greenfield PF (1990) Water Res 24: 583
18. Dochain D, Bastin G (1985) Environ Technol Lett 6: 584
19. Dochain D, Perrier M, Pauss A (1991) Ind Eng Chem Res 30: 129
20. Dolfing J (1988) Acetogenesis. In: Zehnder AJB (ed) Biology of anaerobic microorganisms. Wiley, New York, p 417
21. Erickson LE (1988) Bioenergetics and yields for anaerobic digestion. In: Erickson LE and Fung DY-C (eds) Handbook on anaerobic fermentations. Dekker, New York, p 325
22. Erickson LE, Fung DY-C (1988) Handbook on anaerobic fermentations, Dekker, New York
23. Fan T, Erickson LE, Baltes JC, Shah PS (1973) J Water Pollut Control Fed 45: 591
24. Golden MP, Ydstie BE (1989) AIChE J 35: 1157
25. Gujer W, Zehnder AJB (1983) Water Sci Technol 15: 127
26. Harper SR, Pohland FG (1986) Biotechnol Bioeng 28: 585
27. Hickey RF, Switzenbaum MS (1991) J Water Pollut Control Fed 63: 129

28. Kaspar HF (1977) Untersuchung zur Kopplung von Wasserstoff- und Methanbildung im Faulschlamm. Diss ETH, Zürich
29. Kaspar HF, Wuhrmann K (1978) Appl Environ Microbiol 36: 1
30. Kissalita WS, Lo KV, Pinder KL (1989) Biotechnol Bioeng 33: 623
31. Lloyd D, Whitmore TN (1988) Lett Appl Microbiol 6: 179
32. Märkl H, Mather M, Witty W (1983) Mess- und Regeltechnik bei der anaeroben Abwasserreinigung sowie bei Biogasprozessen. In: Rehm H-J, Reed G (eds) Biotechnology Vol. 2, R. Oldenburg, München, p 369
33. Mather M (1986) Mathematische Modellierung der Methangärung. In: Fortschrittsberichte-VDI, Reihe 14: Landtechnik/Lebensmitteltechnik Vol. 28, VDI Verlag, Düsseldorf
34. McCarty PL (1981) One hundred years of anaerobic treatment. In: Hughes et al. (eds) Anaerobic digestion. Elsevier, Amsterdam, p 1
35. McCarty PL, Smith DP (1986) Environ Sci Technol 20: 1200
36. McCarty PL, Mosey FE (1991) Water Sci Technol 24: 17
37. McInerney MJ (1988) Anaerobic hydrolysis and fermentation of fats and proteins. In: Zehnder AJB (ed) Biology of anaerobic microorganisms. Wiley, New York, p 373
38. Moletta R (1989) Environ Technol Lett 10: 173
39. Mosey FE (1983) Water Sci Technol 15: 209
40. Pauss A, Beauchemin C, Samson R, Guiot S (1990) Biotechnol Bioeng 35: 492
41. Pavlostathis SG, Giraldo-Gomez E (1991) Water Sci Technol 24: 1519
42. Petersen JN, Whyatt GA (1990) Biotechnol Bioeng 35: 114
43. Di Pinto AC, Limoni N, Passino R, Rozzi A, Tomei MC (1990) Anaerobic process control by automated bicarbonate monitoring. In: Briggs R (ed) Instrumentation, control and automation of water and wastewater treatment and transport systems. Pergamon, Oxford, p 51
44. Powell GE, Archer DB (1989) Biotechnol Bioeng 33: 570
45. Da Pra E (1987) Rechnerunterstützte Optimierung des Abbaus von Rübenschwemmwasser mittels eines Anaerobfilters Diss. Univ, Zürich
46. Renard P, Dochain D, Bastin G, Naveau H, Nyns EJ (1988) Biotechnol Bioeng 31: 287
47. Ripley LE, Boyle WC, Converse JC (1986) J Water Pollut Control Fed 58: 406
48. Roels JA (1983) Energetics and kinetics in biotechnology. Elsevier Biomedical Press, Amsterdam
49. Rolf MJ, Lim HC (1984) Chem Eng Commun 29: 1236
50. Rozzi A, Eng M (1984) Trans Inst M C 6: 153
51. Rozzi A, Di Pinto AC, Brunetti A (1985) Environ Technol Lett 6: 594
52. Ryhiner G (1990) Regelung und adaptive On-Line Optimierung von ein- und zweistufigen anaeroben Fliessbett-Reaktoren. Diss ETH, Zürich
53. Ryhiner G, Dunn IJ, Heinzle E, Rohani S (1992) J Biotechnol 22: 89
54. Bélaich JP, Bruschi M, Garcia JL (1990) Microbiology and biochemistry of strict anaerobics involved in interspecies hydrogen transfer. FEMS Symp. No. 54. Plenum New York
55. Schürbüscher D, Wandrey Ch (1991) Anaerobic waste water process models. In: Rehm HJ, Reed G (eds) Biotechnology Vol. 4, VCH, Weinheim, p 441
56. Shi Z, Shimizu K, Watanabe N, Kobayashi T (1989) Biotechnol Bioeng 33: 999
57. Smith DP, McCarty PL (1990) J Water Pollut Control Fed 62: 175
58. Stephanopoulos G (1988) Modelling mixed culture interactions in anaerobic digestion. In: Erickson LE and Fung DY-C (eds) Handbook on anaerobic fermentations. Dekker, New York, p 597
59. Stouthamer AH (1988) Bioenergetics and yields with electron acceptors other than oxygen. In: Erickson LE and Fung DY-C (eds) Handbook on anaerobic fermentations. Dekker, New York, p 345
60. Switzenbaum MS, Giraldo-Gomez E, Hickey RF (1990) Enzyme Microb Technol 12: 722
61. Thauer, Jungermann, Decker (1977) Bacteriol Rev 41: 147
62. Tschui M (1989) Dynamisches Verhalten der mesophilen anaeroben Schlammstabilisierung. Dissertation. ETH, Zürich
63. Vogels GD, Keltjens JT, Van der Drift C (1988) Biochemistry of methane production. In: Zehnder AJB (ed) Biology of anaerobic microorganisms. Wiley, New York, p 707
64. Wandrey C, Aivasidis A, Bastin KH (1986) Ann NY Acad Sci 469: 421
65. Wang YT, Suidam MT, Rittman BE (1986) J Environm Eng 112: 975
66. Whitmore TN, Lloyd D (1986) Biotechnol Lett 8: 203
67. Whitmore TN, Lazzari M, Lloyd D (1985) Biotechnol Lett 7: 203

68. Whitmore TN, Jones G, Lazzari M, Lloyd D (1987) Methanogenesis in mesophilic and thermophilic anaerobic digesters: Monitoring and control based on dissolved hydrogen. In: Heinzle E, Reuss M (eds) Mass spectrometry in biotechnological process analysis and control. Plenum, New York, p 143
69. Wiesmann U (1988) Chem Ing Tech 60: 464
70. Zehnder AJB (1988) Biology of anaerobic microorganisms. Wiley, New York
71. Ricker NL, Slater WR, Merigh M (1988) Process monitoring and control strategies for anaerobic wastewater treatment. DECHEMA Biotechnol Conf Vol. 2, p 413
72. Costello DJ, Lee PL, Greenfield PF (1989) Bioprocess Eng 4: 119

Design of Reactors for Plant Cells and Organs

Pauline M. Doran
Department of Biotechnology, University of New South Wales,
P.O. Box 1, Kensington NSW 2033, Australia

Advances in Biochemical Engineering
Biotechnology, Vol. 48
Managing Editor: A. Fiechter

To be economically feasible, production of most secondary metabolites by plant-cell culture requires high cell densities. In standard reactor configurations such as stirred-tank and airlift vessels, large-scale culture of plant cells encounters problems of shear damage, poor mixing, and inadequate mass transfer of oxygen. Immobilised cells and organised tissue such as embryos, roots and shoots often give enhanced levels of production; however, design of appropriate reactors is at an early stage. This paper identifies the rate-limiting processes in plant-cell culture, and discusses ways for improving reactor performance.

1 Introduction

Cultured plant cells are capable of producing a range of valuable secondary metabolites. Research into large-scale plant-tissue culture is aimed mainly at commercial manufacture of these compounds, although reactor culture of organised tissue can also be used for plant propagation. The shikonin process developed by Mitsui Petrochemical Industry Ltd in Japan demonstrated in 1983 that industrial-scale culture of dispersed plant-cells is technologically feasible. A major effort is still being made to define the requirements of suspended plant-cell cultures and their responses to different reactor conditions. This has proven to be a difficult and challenging area of engineering research. Familiar problems of reactor hydrodynamics, mixing, and oxygen transfer are coupled with plant-cell shear-sensitivity and aggregation, formation of non-Newtonian suspensions at relatively low cell densities, and poor definition of conditions required for secondary synthesis. Although reactors up to 20000 litres have been used to culture plant cells, the task of designing an appropriate reactor system for secondary-metabolite production has not been completed. Yields of most secondary compounds in vitro are still very low and decline even further after scale-up. In most cases, economics demands that high cell densities be used to raise volumetric productivity; a reactor capable of supporting 100 kg m^{-3} dry weight suspended plant-cells without mixing, mass transfer or shear-damage problems is yet to be found. While immobilised cells and differentiated plant tissue are rapidly replacing suspended plant-cells as the focus of research, recent progress with enzyme regulation and genetic manipulation of plant cells may eventually improve the commercial prospects of large-scale suspended-cell culture.

Interest in immobilising plant cells followed from the theory that secondary metabolites are more likely to be produced by slow-growing cells with high cell-cell contact than by dispersed suspended cells. Early work showed that product yield and excretion could be greatly improved by immobilisation, with the added benefit of reduced shear damage to cells. This promoted several studies of immobilisation methods and bioreactors suitable for large-scale operations. The natural tendency of plant cells to clump together is being exploited; self-immobilised or aggregated plant-cells can be grown in packed- and fluidised bed reactors. Surface-attachment of plant cells in membrane and biofilm reactors is also being investigated. Provision of adequate oxygen transfer and maintenance of aseptic conditions during reactor operation are two important factors in immobilised plant-cell reactor design.

Once it was demonstrated that hairy-root cultures produce high levels of secondary metabolites there was a need for appropriate reactors. Work in this new area of bioreactor design is progressing rapidly. A major problem with large-scale culture of roots is poor distribution of biomass. Novel systems are now being developed to fix the roots in place so the vessel can be completely filled with tissue. In addition, spray reactors are being used instead of submerged culture; this mode of culture promotes oxygen transfer and provides low-shear conditions. Another innovation is use of bioreactors to grow somatic embryos; this technology could be applied for mass propagation of plants and production of artificial seed as well as for large-scale phytochemical synthesis. Differentiated shoots and plantlets have recently been cultivated in reactors. New ideas for providing light to photoautotrophic plant cells in large-scale vessels are also now appearing.

The aim of this review is to identify problems associated with large-scale culture of various forms of plant tissue and to summarise the most recent developments in reactor technology. Emphasis is given to the engineering aspects of plant culture. More general treatment of applied plant-tissue culture can be found in previous articles by Fowler [1], Misawa [2], Kurz and Constabel [3], and Rhodes et al. [4].

2 Suspended Plant Cells

2.1 *Properties of Plant Cultures Relevant to Reactor Design*

Most of the technological challenges associated with large-scale culture of suspended plant cells can be traced to certain biological characteristics. High-density plant-cell slurries are difficult to mix and keep in suspension, resulting in poor distribution of oxygen and nutrients in reactors. Non-Newtonian rheology, tendency to aggregate and shear sensitivity compound these problems. Details of these parameters and their influence on reactor design are described in the following sections.

2.1.1 Rheology

Plant culture rheology remains poorly characterised. The morphology of plant cells in vitro is highly varied; near-spherical and elongated cells co-exist in suspension cultures, and there is usually a wide size-distribution. These properties mean that high-density plant-cell suspensions become viscous, to the order of several hundred centipoise. High viscosity leads to longer circulation times in reactors, and formation of dead zones where the liquid remains stagnant. Reduced turbulence in viscous suspensions has a pronounced effect on oxygen transfer as air bubbles are not dispersed as easily as in non-viscous media.

Kato et al. [5] determined that *Nicotiana tabacum* cultures acted as non-Newtonian fluids even at low cell concentrations of 0.9 kg m^{-3}. The culture filtrate after removal of cells maintained a Newtonian character through the entire culture period. As the cell concentration grew from 0.9 to 13 kg m^{-3} dry weight, filtrate viscosity increased from 0.9 mPa s to only 2.2 mPa s even though apparent viscosity of the whole cell suspension increased by a factor of 27.5. This small increase in filtrate viscosity indicates that excretion of extracellular material by plant cells does not contribute significantly to the overall viscosity; the viscous and non-Newtonian character of the suspension is due mainly to the solids content.

Several models are available to describe non-Newtonian fluids; those applicable to microbial cultures are discussed by Roels et al. [6]. Bingham and power-law models have been used to describe plant-cell suspensions. The equations are:

Bingham fluid: $$\tau = \tau_0 + K\dot{\gamma}^n \qquad (1)$$

Power-law fluid: $$\tau = K\dot{\gamma}^n \qquad (2)$$

where τ is shear stress, τ_0 is the yield stress, K is the consistency index, $\dot{\gamma}$ is shear rate, and n is the flow behaviour index.

By analogy with Newton's law of viscosity, an apparent viscosity μ_{app} can be defined from Eq. (2) for power-law fluids:

$$\mu_{app} = K\dot{\gamma}^{n-1} \quad (3)$$

so that:

$$\tau = \mu_{app}\dot{\gamma} \,. \quad (4)$$

Unlike Newtonian fluids, the apparent viscosity of plant-cell cultures depends on shear intensity.

Measurement of empirical constants such as τ_0, K and n is difficult for fluids containing solids. Practical limitations associated with viscosity measurements are discussed by Charles [7] and Metz et al. [8]. For suspensions, these include destruction of pellets and flocs during analysis, settling of solids, and formation of less dense layers at the walls of the measuring device. Some of these problems can be alleviated by use of a turbine impeller for viscometry studies rather than conventional instruments such as the rotating cylinder [6].

Wagner and Vogelmann [9], Kato et al. [5], Tanaka [10] and Scragg et al. [11] have demonstrated that plant-cell cultures are pseudoplastic or shear-thinning at shear rates up to $1000\,s^{-1}$, so that n in Eqs. (1)–(3) is <1. Time-dependent thixotropic behaviour has been reported [9, 11]; this is probably due to progressive break-up of cell aggregates when shear levels are maintained. Non-Newtonian behaviour involving a yield stress (τ_0) has been shown to occur in *Morinda citrifolia* [9] and *Catharanthus roseus* [12] cultures; however, Scragg et al. [11] report that evidence of a yield stress for *Catharanthus roseus* is dependent on the measurement technique employed.

Shear thinning in plant-cell cultures means that apparent viscosity is lower in areas of high shear, e.g. in the impeller region of stirred reactors. Air bubbles will rise rapidly through these regions; away from the impeller dead zones will occur more readily as apparent viscosity increases. If yield stress is a feature of plant-cell rheology, this may have implications for aeration. Small bubbles rising in the culture liquid may not exert a sufficiently high stress on the surrounding fluid and will remain fixed in the same fluid for long periods of time. Bubbles circulating with the fluid rather than following the rising path of larger bubbles will become depleted of oxygen, so that values for total gas hold-up in the reactor will give a distorted picture of mass-transfer conditions.

In some cases viscosity models such as Eqs. (1) and (2) are valid only under limited conditions; K and n values may vary with sufficiently large changes in shear intensity. Because the rheological behaviour of plant-cell cultures is of critical importance in reactor design, investigation of these parameters is required. The limited information currently available is summarised in Table 1.

In studies conducted so far, the power-law index n for plant cells has remained largely independent of aggregate size and cell concentration. The consistency index

Table 1. Rheological parameters for plant-cell suspensions

Species	n	τ_0 (Pa)	Reference
Nicotiana tabacum	0.69–0.74	–	[5]
	0.70	–	[10]
	0.73	–	[25]
Vinca rosea	0.53	–	[10]
Cudrania tricuspidata	0.53	–	[10]
Licorice	0.7	–	[25]
Morinda citrifolia	–	2–4	[9]
Catharanthus roseus	–	300	[12]

K on the other hand is strongly dependent on cell concentration. Absolute values of K have not been reported for plant cells; however, for biomass levels between 10 and 15 kg m^{-3} dry weight, Tanaka [10] showed that K values for several plant species varied in proportion to cell concentration raised to a power between 6 and 7. This relationship is similar to that found for pelleted cells of the mould *Paecilomyces varioti* [13], although there is a general lack of consensus about the relationship between biomass concentration and viscosity for mycelial cultures [8]. If K is so strongly dependent on plant-cell concentration, apparent viscosity of plant-cell suspensions can increase tremendously over the course of batch culture.

Few researchers have reported operating viscosities in plant-cell reactors. Scragg et al. [11] estimate the apparent viscosity of *Catharanthus roseus* cells at a concentration of 250 kg m^{-3} fresh weight (which would correspond to 10–15 kg m^{-3} dry weight) and a shear rate of 609 s^{-1} as approx. 3 mPa s. Significantly greater apparent viscosities can be expected at higher cell densities and at the lower shear-rates found in airlift and stirred reactors. Even so, maximum values would be an order of magnitude less than those achieved in many mycelial cultures.

2.1.2 Aggregate Formation

Aggregation is generally a poorly-controlled parameter in plant-cell suspensions; aggregates and flocs range in size from two cells to clumps of 0.5 cm diameter or more. Wagner and Vogelmann [9] observed that plant-cell cultures forming flocs or pellets caused fewer macromixing problems and less shear damage than suspensions containing mostly single cells. Tendency towards clumping and wall growth is somewhat species dependent; reducing the Ca^{2+} level in the medium [14, 15] and addition of sorbitol and cell-wall-degrading enzymes such as pectinase and cellulase [16] have been reported to reduce aggregate formation. In bioreactors, the size of aggregates depends also on the shear levels present [17]. Control of clumping may be achieved to some extent by changing reactor geometry and operating conditions.

The range of aggregate size and specific gravity in several plant cultures is shown in Table 2. These data can be compared with the density of the suspending liquid. Murashige and Skoog medium containing 3% sucrose has a specific gravity of about 1.013; this reduces to about 1.000 at the end of batch culture [10].

Table 2. Size and specific gravity of plant-cell aggregates in suspension culture. (Adapted from Tanaka [10])

Species	Aggregate size (μm)	Aggregate specific gravity
Nicotiana tabacum	150–800	1.005–1.015
Vinca rosea	44–500	1.002–1.005
Agrostemma githago	44–2000	1.015–1.026
Cudrania tricuspidata	44–2000	1.018–1.023

2.1.3 Shear Sensitivity

Data on plant-cell shear sensitivity show a wide range of responses. Wagner and Vogelmann [9] reported that *Catharanthus roseus* cells were completely destroyed within 5 days at a turbine-impeller speed of 28 rpm; based on the correlation of Metzner and Otto [18] this corresponds to an average shear rate of only $5\ s^{-1}$. In more recent studies, Hooker et al. [19] found extreme shear damage in *Nicotiana tabacum* cultures in a 3-litre reactor with a flat-blade impeller operated at 200 rpm. Scragg et al. [20] tested the response of *Catharanthus roseus* to shear in a 3-litre stirred tank operated at 1000 rpm. The average shear rate was $167\ s^{-1}$ corresponding to a maximum shear rate of about $1500\ s^{-1}$ at the impeller tip. At the end of 5 h, 30% of the cells had been destroyed; the remainder were viable and survived subsequent culture [11].

As well as mechanical disruption and loss of viability, there are other more subtle effects of shear stress on plant cells. A study of *Catharanthus roseus* showed that hemicellulose, cellulose and pectin levels in the cell wall depended on the level of hydrodynamic stress imposed during culture [17]. The mass ratio of cell wall to whole cell also increased with intensity of shear. Plant cells are reported to be more susceptible to shear during late exponential- and early stationary-phases when the cells are of relatively large size and contain large vacuoles [9, 21].

Variation in resistance to shear between plant species and the ability of cultured cells to develop shear tolerance raise questions about whether plant cells are always shear sensitive. Scragg et al. [11] and Allan et al. [22] report that cells initially shear sensitive developed shear tolerance after 2–5 years cultivation in vitro. Most studies of shear sensitivity in plant-cell cultures have been carried out over short periods of time relative to the duration of batch culture. However, long-term shear responses for four plant species: *Catharanthus roseus, Nicotiana tabacum, Cinchona robusta* and *Tabernaemontana divaricata*, have been tested recently over 10–14 d in a 3-litre reactor with a 6-blade Rushton impeller [23]. The stirrer was operated at 250 and 1000 rpm. *C. roseus* and *N. tabacum* were shear resistant under these conditions; both *C. robusta* and *T. divaricata* suffered growth impairment and cell disruption above 250 rpm. Shear tolerance was correlated in this study with stability of growth characteristics and time since initiation of culture.

Sensitivity to shear has been determined in most cases by varying the shear rate, $\dot{\gamma}$. In work by Midler and Finn [24] on disruption of protozoa, cell suspensions of high apparent viscosity suffered more damage at constant shear rate than suspensions with low apparent viscosity. Accordingly, shear stress rather than shear rate appears to be the important variable in assessing shear sensitivity of cells. When high local velocity-gradients are present, such as at the tip of the impeller in stirred systems, the extent of cell damage may also depend on the maximum rather than average shear stress. Maximum shear stresses withstood by plant cells in laminar flow in a high-shear Couette rheometer have been reported by Chen et al. [25]. For tobacco cells the critical shear level was 2.5 Pa and for licorice cells 8 Pa. These figures indicate a significantly lower tolerance to shear stress than that observed by Meijer [23] with plant cells subjected to stirrer speeds of 250–1000 rpm.

The actual mechanism of shear damage to plant cells has not been studied. However, hydrodynamic effects on animal cells have been analysed by several groups [26–34]; the results may also be applicable to suspended plant-cells. A complete picture of shear effects in stirred and aerated reactors has not yet been determined unequivocally. In stirred microcarrier-cultures of animal cells where there is no entrainment of gas bubbles, interactions between cells and turbulent eddies are considered most likely to cause shear damage [26, 35]. If the size of the eddies is approximately the size of the microcarriers, the eddies dissipate their energy against the surface of the microcarriers and cause damage to the attached cells. The dimension of the smallest eddies in turbulent flow is given by the Kolmogorov scale:

$$\eta = \left(\frac{\nu^3}{\varepsilon}\right)^{1/4} \tag{5}$$

where η is eddy size, ν is fluid kinematic viscosity, and ε is local energy dissipation rate per unit mass of liquid. In large-scale equipment with low-viscosity liquids, η is of the order 30–100 μm; smaller eddies are often produced in laboratory-scale vessels. A relationship between cell damage in microcarrier cultures and η has been demonstrated; decreasing the Kolmogorov scale below $^1/_2$–$^2/_3$ the microcarrier diameter increases cell damage [26, 28]. Croughan et al. [28] also showed that increased viscosity, which results in increased Kolmogorov scale, reduces damage. Since plant cells have dimensions in the vicinity of 30–100 μm and can form clumps of even greater size, it is likely that interaction with turbulent eddies also causes shear damage in plant suspensions. To date, however, there has been no experimental investigation of these effects.

In aerated systems, shear damage can occur at much lower impeller speeds than those producing eddies of the dimensions described above. There is evidence that in systems with bubble entrainment, significant shear damage occurs at the liquid surface as a result of bubble disengagement [32]. Data of Handa-Corrigan et al. [29] relating cell viability to bubble-column height and formation of a stable protective foam support this theory. Investigations by Kunas and Papoutsakis [30] have distinguished shear-damage mechanisms in reactors with and without bubble entrainment. At low agitation speeds between 150 rpm and 600 rpm, damage to hybridoma cells was insignificant if there were no gas-liquid interface in the vessel. Damage at these stirrer speeds occurred only when bubbles were allowed to burst at the liquid surface. At 800 rpm, which corresponded to a Kolmogorov-eddy size comparable with the size of the cells, damage to the cells increased but was still less significant in the absence of a gas interface.

Studies of shear damage in plant-cell suspensions have concentrated exclusively on the effects of mechanical agitation. While different species can be expected to have different shear sensitivities, the extremely wide variation in shear response from complete cell destruction at 28 rpm [9] to tolerance of 1000 rpm [23] may be due to intrusion of other factors not considered in these studies. The shear stress produced by bursting bubbles may be one such factor.

2.2 Stirred Versus Airlift Reactors

Much of the literature on plant-cell reactors has focused on experimental comparison of airlift and stirred-tank designs [9, 10, 36–38]. In early work, Wagner and Vogelmann [9] measured cell growth and alkaloid production by *Catharanthus roseus* and *Morinda citrifolia* in a selection of bioreactors including shake flasks, airlift vessels and stirred tanks with different impeller designs. They concluded that airlift reactors provided the best conditions; mechanically agitated devices caused excessive shear damage. After considering the low oxygen demand of plant cells and predicting that mass transfer requirements could be adequately supplied by pneumatic agitation, Fowler, Scragg and co-workers conducted extensive studies of airlift reactors up to 80-litres capacity for culture of *Catharanthus roseus* [39–42]. Maximum biomass concentration in these trials was about 23 kg m^{-3} dry weight; the reactors were usually operated below 12 kg m^{-3}.

Economic analysis of commercial plant-cell culture shows that high cell densities (up to 100 kg m^{-3}) are required to compensate for the low production rates and yields normally found with dedifferentiated tissue [43]. There are several reports [36, 37, 44] that performance of airlift reactors becomes severely limited by poor gas disengagement, impeded liquid circulation and development of unmixed zones at biomass densities above 20–30 kg m^{-3} dry weight. Increasing aeration rates to improve mixing has not been successful; over-ventilation of cultures reduces levels of dissolved carbon dioxide and other gaseous components such as ethylene, so that growth and production declines [39, 40]. These apparent shortcomings of airlift reactors have re-awakened interest in stirred reactors for plant tissue culture. Tobacco cells have been grown successfully in a 20000-litre stirred tank [45], while Mitsui Petrochemical Industries Ltd. uses two stirred reactors of 200-litres and 750-litres capacity for commercial production of shikonin [46]. Stirred vessels tested by Mitsui for industrial production of berberine are able to support *Coptis japonica* cell densities of 70 kg m^{-3} dry weight [47]. In 1986 a German company DIVERSA set up a cascade of five stirred tanks in series to study the economic aspects of large-scale plant-cell culture for secondary metabolite production [48]. Reassessment of stirred reactors has also been encouraged by recent findings that cultivated plant cells are able to develop resistance to shear [22, 49].

Because of their widespread application, airlift and stirred-tank reactors have been the subject of extensive chemical-engineering analysis. Equations developed for reactor design should be applicable to plant-cell cultures and provide information about the effects of changing culture conditions on reactor performance. Predictive equations for stirred tanks containing non-Newtonian fluids were developed 15–20 years ago; hydrodynamics, mixing and mass transfer in airlift reactors have been studied much more recently. The following section summarises available design information for stirred and airlift vessels; several of the equations presented are applied in a theoretical analysis of these devices for large-scale culture of suspended plant cells.

2.3 Engineering Analysis of Reactors

Figure 1 shows standard configurations of airlift and stirred-tank reactors.

In stirred-tank reactors, mixing and dispersion of air is achieved by mechanical agitation; this requires a relatively high energy input per unit volume which is ultimately dissipated as heat. High mass-transfer rates can be achieved, although for a given mass-transfer coefficient stirred-tank reactors impart higher levels of shear than non-mechanically agitated vessels [36]. For long-term aseptic operation good-quality sterile seals are required where the agitator shaft enters the vessel. Temperature, pH, dissolved oxygen concentration and nutrient levels are easy to control in stirred tanks. Stirred reactors are usually equipped with baffles to reduce

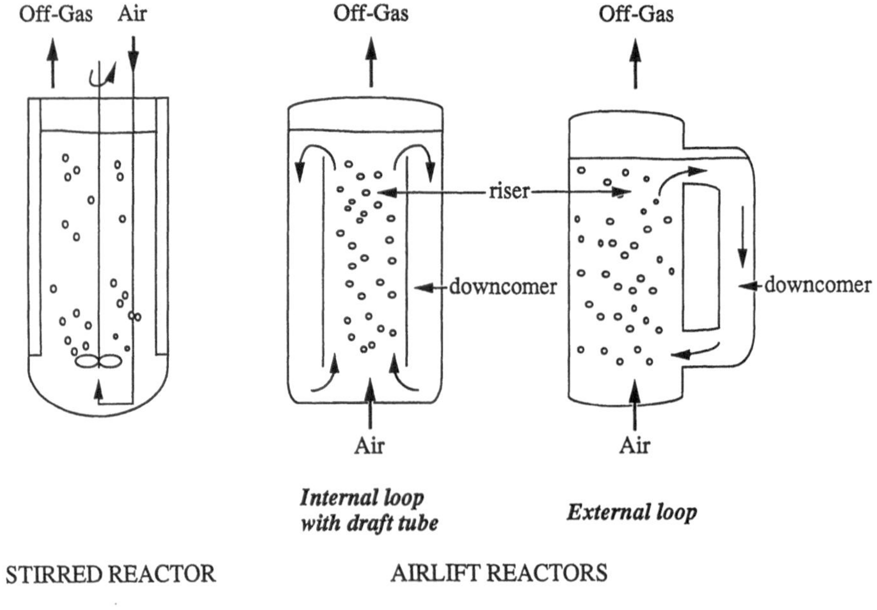

Fig. 1. Stirred-tank and airlift reactor configurations

vortexing; a wide variety of impeller sizes and shapes can be used to produce different flow patterns inside the reactor. In tall vessels, installation of two or more sets of impellers improves mixing.

In airlift reactors, air sparging is used for both mixing and aeration. The reactor is divided into two distinct zones. Gas is sparged into the riser section; the increased gas hold-up and decreased fluid density in this section cause the fluid to rise. Gas disengages at the top of the reactor leaving heavier liquid which recirculates through the downcomer. Low and relatively uniform shear fields, low power input and extended aseptic operation are important advantages of airlift reactors for plant-cell operations.

As shown in Fig. 1 there are two basic types of airlift reactor:

a) the internal-loop airlift in which riser and downcomer are separated by an internal baffle or draft tube; and
b) the external- or outer-loop airlift where riser and downcomer are separate tubes connected by short horizontal sections at top and bottom.

Both types of airlift usually have circular cross-sections. In external-loop reactors the head-space region at the top of the vessel can be modified to change the efficiency of gas-liquid separation; this determines whether bubbles are recirculated or not and can affect gas hold-up, liquid velocity and turbulence. In internal-loop vessels either the draft tube or annulus may be sparged. Changing the distance between the lower edge of the baffle and the bottom of the reactor can affect the pressure drop in this region with consequences for liquid recirculation velocity, gas hold-up and mixing. Gas-liquid separation depends on the depth of submersion of the upper edge of the draft tube in the liquid. In internal-loop vessels, since the downcomer and riser connect directly at the top of the draft tube, gas-liquid separation is generally not as effective as in external-loop devices and gas bubbles are often drawn into the downcomer. The capabilities of internal- and external-loop configurations are compared in Table 3 in terms of various performance indicators [50].

Table 3. Relative performance of internal- and external-loop airlift reactors. (Adapted from Chisti [50])

Parameter	Reactor	
	Internal-loop	External-loop
Mass transfer coefficient	higher	lower
Overall gas hold-up	higher	lower
Gas hold-up in riser	higher	lower
Gas hold-up in downcomer	higher	lower
Superficial liquid velocity in riser	lower	higher
Circulation time	higher	lower
Liquid Reynolds number (shear)	lower	higher
Heat transfer	probably lower	probably higher

For plant-cell suspensions it is crucial that the reactor provides good mixing and suspension of solids, adequate oxygen transfer, and non-damaging levels of hydrodynamic shear.

2.3.1 Mixing

Good mixing achieves homogeneous distribution of oxygen and other nutrients in the bulk fluid; dead zones where anoxia or nutrient starvation can occur are prevented. Mixing also reduces localised build-up of substances such as sugar which can affect cell metabolism if allowed to accumulate in unmixed regions. Both heat and mass transfer are greatly influenced by mixing and the intensity of turbulence in the reactor.

The efficiency of mixing is usually quantified in terms of mixing time. Mixing time (t_{mx}) is defined as the time required to reach some arbitrary level of uniformity in the liquid being mixed. In a recirculating-flow mixed reactor, the number of circulations required for complete mixing depends on the degree of turbulent diffusion. For a stirred tank with several baffles and small impeller, a satisfactory relationship between circulation time and mixing time is [51]:

$$t_{mx} = 4t_c \tag{6}$$

where t_c is the time taken for liquid to travel one complete circulation loop in the reactor. For internal-loop airlift reactors where gas is sparged into the annulus:

$$t_{mx} = 3.5t_c \left(\frac{A_d}{A_r}\right)^{0.5} \tag{7}$$

and for external-loop airlifts:

$$t_{mx} = 5.2t_c \left(\frac{A_d}{A_r}\right)^{0.5} \tag{8}$$

where A_d is the downcomer cross-sectional area, and A_r is the riser cross-sectional area [52].

For the same total power input per unit volume (P/V_L) specific mixing times (t_{mx}/V_L) for water in either internal- or external-loop airlifts are three to five times longer than in stirred tanks [52]. The mechanisms of mixing in stirred tanks and airlift reactors are different and are reviewed separately.

2.3.1.1 Stirred Tanks

A wide range of mixing equipment is available for mechanical agitation; classification is usually based on liquid viscosity. Figure 2 shows the recommended viscosity ranges for a number of impeller designs [53]. Some of these designs are illustrated schematically in Fig. 3; the most widely studied impeller is the turbine disc with 6 flat blades pitched at 90°.

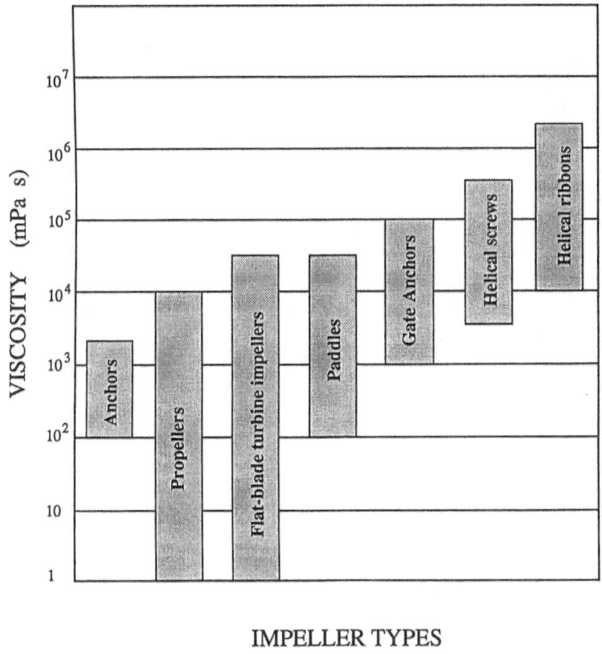

Fig. 2. Viscosity ranges for different impeller designs. (Adapted from Holland and Chapman [53])

Metzner and Otto [18] determined that average shear rate ($\dot{\gamma}_{av}$) in a stirred tank is directly proportional to agitation speed:

$$\dot{\gamma}_{av} = kN_i \tag{9}$$

where the value of k depends on impeller design and type of fluid, and N_i is rotational speed. This relationship has been tested for both Newtonian and non-Newtonian fluids.

Flow visualisation studies with non-Newtonian fluids have shown that flow patterns are different from those set up by the same equipment in Newtonian liquids [54]. For pseudoplastic fluids, apparent viscosity in the impeller region is rather low because of the higher shear rates; viscosity increases with distance from the impeller. The recommended geometry for mixing operations is also different with non-Newtonian fluids. For an ungassed low-viscosity fluid the optimum ratio of tank diameter to impeller diameter ($D_T:D_i$) is 2.5–3.0:1. Reduction of this ratio to 1.5–2.0:1 [55, 56] or use of a close-clearance anchor or helical impeller gives better mixing with viscous non-Newtonian fluids. At low viscosities the volume of fluid set in motion by the impeller is of the order of several cubic impeller-diameters. In viscous fluids this volume decreases markedly. The thickness of the fluid layer moved by the impeller may become only a fraction of the

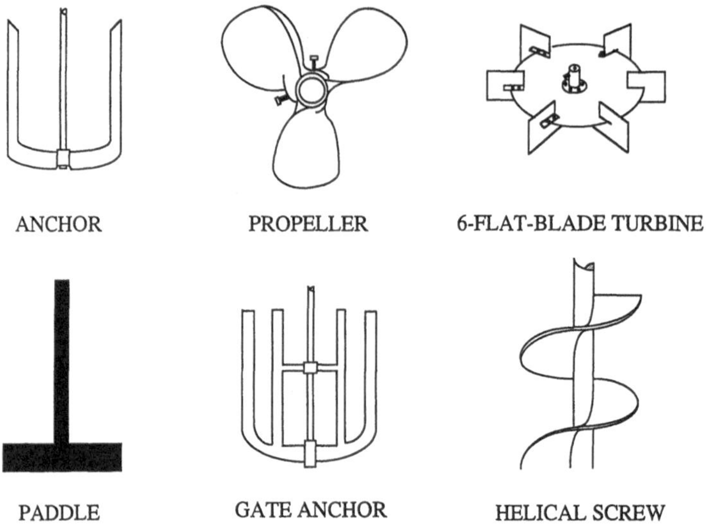

Fig. 3. Common impeller designs for stirred tanks

impeller diameter, so that impellers of small diameter are not likely to provide adequate mixing.

Mixing time in non-aerated Newtonian fluids in baffled cylindrical tanks has been correlated with system geometry and operating parameters by Norwood and Metzner [57]. Flat-blade turbine impellers were used. The dimensionless group, f_m:

$$f_m = \frac{t_{mx}(N_i D_i^2)^{2/3} g^{1/6} D_i^{1/2}}{H_L^{1/2} D_T^{3/2}} \tag{10}$$

was found to be approximately constant and equal to 6 above a Reynolds number (Re_i) of 1.8×10^5, where:

$$Re_i = \frac{D_i^2 N_i \varrho_L}{\mu_L}. \tag{11}$$

In Eqs. (10) and (11) t_{mx} is mixing time; N_i is impeller rotational speed; D_i is impeller diameter, g is gravitational acceleration; H_L is liquid height; D_T is tank diameter; ϱ_L is fluid density; and μ_L is fluid viscosity.

Norwood and Metzner [57] proposed that the relationship between f_m and Re_i for Newtonian liquids could be used to calculate approximate mixing times for pseudoplastic fluids at high Reynolds numbers. For non-Newtonian fluids and viscous cultures of *Streptomyces niveus*, Wang and Fewkes [58] found that mixing times measured in stirred tanks were in good qualitative agreement with the Norwood-Metzner correlation, although the value of f_m at Reynolds numbers above 10^4 was approx. 20. The discrepancy may be due to aeration effects; aeration of suspensions increases mixing times for a given Reynolds number [59]. In addition, Wang and Fewkes calculated the non-Newtonian impeller Reynolds number using the equation proposed by Calderbank and Moo-Young [60]:

$$Re_i = \frac{D_i^2 N_i^{2-n} \varrho_L}{0.1K} \left(\frac{n}{6n + 2}\right)^n \tag{12}$$

where K is the consistency index and n is the flow behaviour index for a power-law fluid. This change should not, however, affect the correlation at high Re_i. Application of the Norwood-Metzner correlation to non-Newtonian cell suspensions is discussed further by Charles [7].

2.3.1.2 Airlift Reactors

Mixing time in airlift reactors depends on the liquid circulation velocity. Higher circulation rates are obtained in external-loop devices, because gas entrainment in the downcomer is greater in internal-loop vessels. Liquid velocities are also dependent on geometric parameters. For internal loops, minimal mixing times have been reported for $0.6 < D_r/D_c < 1.0$, where D_r is riser (draft-tube) diameter and D_c is total column diameter. Equal riser and downcomer cross-sectional area, i.e. $D_r/D_c = 1/\sqrt{2} = 0.71$, is therefore a reasonable design goal [61]. An internal-loop airlift reactor built to these specifications has been used to culture plant cells up to 25 kg m^{-3} dry weight without development of stagnant zones [37]. Mixing times also vary considerably as distance between the top of the draft tube and the liquid level is altered; Weiland [62] measured a three-fold increase in mixing time as this distance was decreased from 0.5 m. Sparger positioning for optimal gas-bubble distribution and liquid circulation has been investigated by Chisti and Moo-Young [63]; Fig. 4 shows proper sparger positioning for both internal- and external-loop devices. The cross-hatched areas in Fig. 4 indicate zones of the reactor which, when filled in, improve liquid flow and prevent biomass settling. Frictional loss at the bottom of internal-loop reactors depends on the area for flow between the riser and downcomer; the effect on liquid velocity of varying geometry in this region of the vessel has been analysed by Chisti et al. [64]. Mixing in airlift reactors can sometimes be improved considerably by attention to geometric details.

Liquid circulation in airlift contactors declines with increasing viscosity [65]. For plant cells, increasing viscosity during batch culture can produce substantial changes in mixing conditions. Although several studies of circulation rates in airlift reactors have been made at low viscosities there are a limited number of correlations for non-Newtonian fluids. Popovic and Robinson [66] developed a relationship between liquid velocity, gas flow rate, reactor geometry and fluid properties for heterogeneous flow in external-loop airlifts at superficial gas velocities ≥ 0.04 m s^{-1}:

$$u_{Lr} = 0.23 u_{Gr}^{0.32} \left(\frac{A_d}{A_r}\right)^{0.97} \mu_{app}^{-0.39} \tag{13}$$

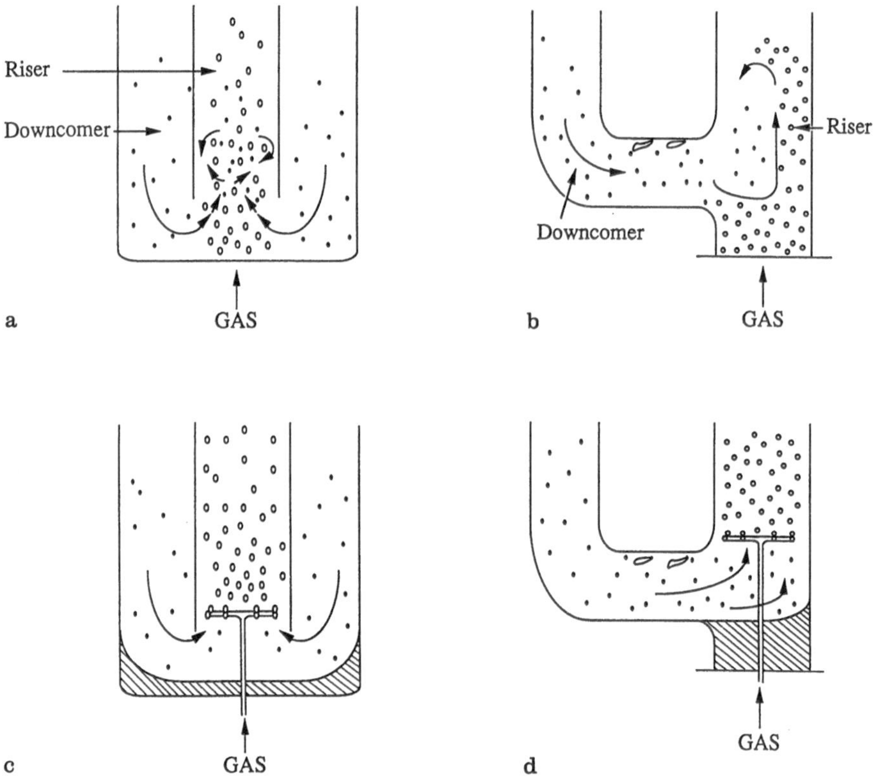

Fig. 4a–d. Influence of sparger location on gas distribution in airlift reactors. Poor distribution of gas in **a)** internal- and **b)** external-loop reactors. Proper positioning of the sparger is shown in **c)** and **d)**. Cross-hatched areas indicate zones which should be filled in to improve liquid flow and prevent biomass settling. (Adapted from Chisti and Moo-Young [63])

where u_{Lr} is superficial liquid velocity in the riser, m s^{-1}; u_{Gr} is superficial gas velocity in the riser, m s^{-1}; A_d is downcomer cross-sectional area, m^2; A_r is riser cross-sectional area, m^2; and μ_{app} is apparent viscosity, Pa s. Equation (13) was developed using viscous pseudoplastic suspensions of 0.5–1.5 wt% carboxymethyl cellulose (CMC) in 0.5 M Na_2SO_4 in external-loop devices with A_d/A_r ratios of 0.111, 0.250 and 0.444. The liquid height in these reactors was 1.88 m. The apparent viscosity was estimated using the correlation for shear rate in bubble columns reported by Nishikawa et al. [67]:

$$\dot{\gamma} = 5000 u_G \tag{14}$$

where u_G is superficial gas velocity, m s^{-1}; and $\dot{\gamma}$ is shear rate, s^{-1}. This relationship was originally developed for bubble columns with CMC suspensions and

$u_G \geq 0.04$ m s^{-1}. From Eq. (14) the value of μ_{app} in Eq. (13) for pseudoplastic fluids can be taken as:

$$\mu_{app} = K(5000u_{Gr})^{n-1} \tag{15}$$

where μ_{app} is apparent viscosity, Pa s; u_{Gr} is superficial gas velocity in the riser, m s^{-1}; n is the flow behaviour index; and K is the fluid consistency index, Pa s^n. When applied to *Aspergillus niger* suspensions containing up to 20 kg dry weight per m^3 cells, Eq. (13) over-predicted u_{Lr} especially at higher u_{Gr} values [68].

Equation (13) was developed for a specified reactor height; however, since liquid circulation relies on the difference in hydrostatic pressures between riser and downcomer, liquid velocity also depends on the height of the reactor. Most reported data on the effects of reactor height are derived from small equipment; these are difficult to extrapolate because of the predominance of end effects. Rousseau and Bu'Lock [61] determined for $0.68 \leq H_L(\mathrm{m}) \leq 1.64$ that:

$$t_{mx} \propto H_L^{1.7} \tag{16}$$

where H_L is liquid height in the reactor, m. This relationship was developed using air-water in an internal-loop device with D_r/D_c between 0.2 and 0.6.

Other equations for liquid velocity have been developed from theoretical analysis of the hydrodynamics in external- and internal-loop airlift reactors [50]. An energy balance for non-Newtonian liquids leads to the equation:

$$u_{Lr}A_r(\Delta P_{Fr} + \Delta P_{Fd}) - \varrho_L g H_D u_{Lr} A_r(\varepsilon_r - \varepsilon_d) + \frac{\varrho_L u_{Lr}^3 A_r}{2}\left[\frac{K_T}{(1-\varepsilon_r)^2} + K_B\left(\frac{A_r}{A_d}\right)^2 \frac{1}{(1-\varepsilon_d)^2}\right] = 0 \tag{17}$$

where u_{Lr} is superficial liquid velocity in the riser, m s^{-1}; ΔP_{Fr} is frictional pressure drop in the riser, Pa; ΔP_{Fd} is frictional pressure drop in the downcomer, Pa; ϱ_L is density of the liquid, kg m^{-3}; g is gravitational acceleration, m s^{-2}; H_D is gas-liquid dispersion height, m; ε_r is gas hold-up in the riser; ε_d is gas hold-up in the downcomer; K_T is frictional loss coefficient for the top of the reactor; K_B is frictional loss coefficient for the bottom of the reactor; A_r is riser cross-sectional area, m^2; and A_d is downcomer cross-sectional area, m^2. To use Eq. (17), gas hold-up in the riser ε_r must be known or estimated from independent correlations, ΔP_{Fr} and ΔP_{Fd} are determined from shear stress calculations, and K_B and K_T are estimated from correlations involving the area for flow between riser and downcomer. Solution for u_{Lr} requires trial-and-error procedures.

Once u_{Lr} is determined, the superficial liquid velocity in the downcomer u_{Ld} can be estimated from continuity:

$$u_{Lr}A_r = u_{Ld}A_d\,. \tag{18}$$

For internal-loop airlift reactors and for external-loop reactors where the lengths of horizontal top and bottom sections are small, circulation time can be determined from u_{Lr}, u_{Ld} and geometric parameters:

$$t_c = \frac{H_r}{u_{Lr}} + \frac{H_d}{u_{Ld}} \tag{19}$$

where H_r is height of the riser and H_d is height of the downcomer. Mixing time can then be calculated from Eq. (7) or (8).

2.3.2 *Mass Transfer*

Of all the nutrients supplied to plant-cell cultures, oxygen is often the most difficult to provide in non-limiting quantities. The critical dissolved-oxygen concentration for plant cells has been reported as $1.3–1.6 \times 10^{-3}$ kg m^{-3} [69, 70]; this is roughly 20% saturation under average culture conditions. Above this level growth is not oxygen-limited and follows first-order kinetics. Below the critical concentration growth is linear with respect to time [71, 72]. Oxygen supply also affects secondary metabolite production [73, 74] and organogenesis [69]; these effects are more difficult to generalise or predict.

The extent to which reactors provide favourable conditions for mass transfer is reflected in the value of the overall liquid-phase mass transfer coefficient, $k_L a_L$:

$$\text{OTR} = k_L a_L (C^* - C_L) \tag{20}$$

where OTR is oxygen transfer rate, kg m^{-3} s^{-1}; k_L is liquid-film mass-transfer coefficient, m s^{-1}; a_L is interfacial area per unit volume of unaerated liquid, m^{-1}; C^* is equilibrium concentration of oxygen in the liquid, kg m^{-3}; and C_L is actual oxygen concentration in the liquid, kg m^{-3}. Equation (20) is valid for local C^* and C_L values; normally C^* and C_L are assumed constant throughout the reactor so gas and liquid phases need to be well mixed. This is acceptable for most low-viscosity liquids but may not be true in high-viscosity systems. In particular, the assumption of well-mixed gas phase is difficult to justify in tall vessels with height:diameter ratios $\gg 1$ [75].

Measured values for mass-transfer coefficients can be expressed as either $k_L a_L$ or $k_L a_D$, where a_D is the gas-liquid interfacial area per unit dispersed volume, m^{-1}. These two mass-transfer coefficients are related by Eq. (21):

$$k_L a_L = \frac{k_L a_D}{1 - \varepsilon_T} \tag{21}$$

where ε_T is total gas hold-up.

For the same total power input per unit volume (P/V_L), much higher $k_L a_L$ values are produced in pneumatically agitated reactors than in stirred tank reactors; this applies also to non-Newtonian fluids [76]. As cell concentration and apparent viscosity increase, the rate of mass transfer can drop to as low as 15% of its initial value [77]. Prediction and improvement of oxygen mass transfer in reactors relies on the prediction and enhancement of $k_L a_L$.

2.3.2.1 Stirred Tanks

In aerated stirred reactors one of the functions of the impeller is to break up and disperse the air bubbles. With pseudoplastic fluids, shear thinning near the impeller causes gas to channel in toward the impeller rather than into the remainder of the tank. This non-uniform distribution is difficult to prevent and has adverse effects on both mass transfer and mixing. Less bubble break-up is observed with viscous non-Newtonian fluids; $k_L a_L$ values tend to be considerably lower than in Newtonian systems under the same operating conditions [78].

Correlations for $k_L a_L$ values in stirred tanks are of two types. Correlations based on power input are reviewed by Schügerl [76] who also discusses effects of gas flow rate, fluid properties and impeller speed on power requirements for non-Newtonian fluids. Other $k_L a_L$ correlations are presented in terms of dimensionless groups representing system geometry, fluid properties and operating conditions.

Rates of oxygen desorption in pseudoplastic non-Newtonian fluids were measured by Yagi and Yoshida [78] in an agitated tank with four baffles and 6-blade turbine impeller. The stirrer speeds tested were 5 s^{-1} to 10 s^{-1}; superficial gas velocities ranged between 0.002 and 0.008 m s^{-1}. The correlation obtained for $k_L a_L$ in non-Newtonian fluids is:

$$\frac{k_L a_L D_i^2}{\mathscr{D}_L} = 0.06 \left(\frac{D_i^2 N_i \varrho_L}{\mu_{app}}\right)^{1.5} \left(\frac{D_i N_i^2}{g}\right)^{0.19} \left(\frac{\mu_{app}}{\varrho_L \mathscr{D}_L}\right)^{0.5} \left(\frac{\mu_{app} u_G}{\sigma}\right)^{0.6} \times \left(\frac{N_i D_i}{u_G}\right)^{0.32} [1 + 2.0 (\lambda N_i)^{0.5}]^{-0.67} \tag{22}$$

where $k_L a_L$ is the mass transfer coefficient, s^{-1}; D_i is impeller diameter, cm; $\mathscr{D}_L$ is diffusivity in the liquid, $cm^2\ s^{-1}$; N_i is rotational speed of the impeller, s^{-1}; ϱ_L is liquid density, g cm^{-3}; μ_{app} is apparent viscosity, g $cm^{-1}\ s^{-1}$; g is gravitational acceleration, cm s^{-2}; u_G is superficial gas velocity, cm s^{-1}; σ is surface tension, g s^{-2}; and λ is characteristic material time, s. For non-Newtonian fluids λ is a constant defined as the reciprocal of the shear rate at which the ratio of apparent viscosity to zero-shear viscosity is 0.67. The correlation of Eq. (22) is limited by determination of λ; viscosity at zero-shear is difficult to measure accurately. Apparent viscosity in Eq. (22) is calculated using the Metzner and Otto [18] relationship for average shear rate as outlined in Eqs. (3) and (9).

An alternative to Eq. (22) is the correlation of Perez and Sandall [79]:

$$\frac{k_L a_L D_i^2}{\mathscr{D}_L} = 21.2 \left(\frac{D_i^2 N_i \varrho_L}{\mu_{app}}\right)^{1.11} \left(\frac{\mu_{app}}{\varrho_L \mathscr{D}_L}\right)^{0.5} \left(\frac{D_i u_G}{\sigma}\right)^{0.447} \left(\frac{\mu_g}{\mu_{app}}\right)^{0.694} \tag{23}$$

where μ_g is gas viscosity, g cm^{-1} s^{-1}; the other symbols and units of Eq. (24) are the same as for Eq. (22). Note that Eq. (23) is a dimensional equation. Apparent viscosity in Eq. (23) is determined by the method of Calderbank and Moo-Young [60]:

$$\mu_{app} = \frac{K}{(11N_i)^{1-n}} \left(\frac{3n+1}{4n}\right)^n . \tag{24}$$

Equation (23) was developed for carbon dioxide absorption in a 0.25% Carbopol solution using a tank equipped with four baffles and a 6-flat-blade turbine impeller operated at $2810 \le Re_i \le 26700$; Re_i is defined by Eq. (11) with the viscosity given by Eq. (24). k_La_L values calculated using Eq. (23) are for carbon dioxide and must be corrected for oxygen transfer. The correction based on the two-film theory of mass transfer is [50]:

$$(k_La_L)_{O_2} = (k_La_L)_{CO_2} \frac{\mathscr{D}_{O_2}}{\mathscr{D}_{CO_2}} \tag{25}$$

where $\mathscr{D}$ is diffusivity. The ratio $\mathscr{D}_{O_2}/\mathscr{D}_{CO_2}$ is equal to approx. 1.2. The applicability of Eq. (23) is discussed further by Yagi and Yoshida [78].

2.3.2.2 Airlift Reactors

Overall mass-transfer coefficients in external- and internal-loop airlift reactors have been studied by a number of investigators. The majority of correlations for k_La_L and gas hold-up involve a simple power-law function of superficial gas velocity:

$$k_La_L \propto u_G^{\beta} . \tag{26}$$

The fundamental hydrodynamic basis for this type of relationship is discussed by Chisti and Moo-Young [80].

Bello et al. [81] found that in internal-loop airlifts gas hold-up in the downcomer was 80-95% that in the riser, while in external-loop reactors downcomer gas hold-up was much lower at 0–50% of riser hold-up. Consequently, the contribution of the downcomer in external-loop devices to oxygen transfer is minimal. Thus, while low downcomer gas hold-up in external-loop airlift reactors promotes liquid circulation, this effect is accompanied by a reduction in oxygen-transfer capability. Complete or near complete gas disengagement in the head-space of external-loop reactors may be harmful to organisms which are sensitive to oxygen concentrations below the critical level. Chisti [50] suggests that in this case the downcomer could be sparged separately with either fresh or recirculated gas.

Gas hold-ups and k_La_L values in airlift contactors decline with increasing apparent viscosity [68, 82]. Presence of suspended solids also has a significant effect on k_La_L due to dampened turbulence, enhanced bubble coalescence and interfacial blanketing; addition of solids above 1% (w/v) has been found to cause an 80% reduction in gas hold-up and oxygen transfer [83].

Popovic and Robinson [84] have proposed a correlation for k_La_D in external-loop airlift reactors:

$$k_La_D = 0.5 \times 10^{-2}\, u_{Gr}^{0.52} \mathscr{D}_L^{0.5} \varrho_L^{1.03} \left(1 + \frac{A_d}{A_r}\right)^{-0.85} \mu_{app}^{-0.89} \sigma^{-0.75} \tag{27}$$

where k_La_D is the mass transfer coefficient based on dispersed volume, s^{-1}; u_{Gr} is superficial gas velocity in the riser, $m\ s^{-1}$; $\mathscr{D}_L$ is the diffusivity of oxygen in the liquid, $m^2\ s^{-1}$; ϱ_L is liquid density, $kg\ m^{-3}$; A_d is downcomer cross-sectional area, m^2; A_r is riser cross-sectional area, m^2; μ_{app} is apparent viscosity, Pa s; and σ is interfacial tension with respect to air, $N\ m^{-1}$. Reactor geometry used to develop Eq. (27) is the same as that described for Eq. (13); apparent viscosity is calculated using Eq. (15). Popovic and Robinson [84] tested Eq. (27) with a range of viscous and non-Newtonian fluids; prediction of k_La_D was within 13% of measured values in each case.

Values of k_La_D from Eq. (27) can be converted to k_La_L using Eq. (21). Gas hold-up in airlift reactors is not uniform; ε_T can be evaluated using Eq. (28):

$$\varepsilon_T = (A_d\varepsilon_d + A_r\varepsilon_r)/(A_d + A_r) \tag{28}$$

where ε_d is gas hold-up in the downcomer and ε_r is gas hold-up in the riser. For external-loop vessels ε_d can be taken as approximately $0.3\varepsilon_r$; ε_r is given by:

$$\varepsilon_r = 0.465 u_{Gr}^{0.65} \left(1 + \frac{A_d}{A_r}\right)^{-1.06} \mu_{app}^{-0.103} . \tag{29}$$

Equation (29) was developed by Popovic and Robinson [84] for viscous CMC solutions, but has been shown by Allen and Robinson [68] to apply to *Aspergillus niger* suspensions at $0.02 \leq u_{Gr}\ (m\ s^{-1}) \leq 0.2$ and $0.01 \leq \mu_{app}\ (Pa\ s) \leq 0.5$.

An alternative equation for k_La_L in external-loop airlift reactors has been proposed by Chisti et al. [83]. This correlation involves solids concentration explicitly:

$$k_La_L = \left(1 + \frac{A_d}{A_r}\right)^{-1} (0.349 - 0.102 C_s)\, u_{Gr}^{0.837 \pm 0.062} \tag{30}$$

where k_La_L is the mass transfer coefficient, s^{-1}; A_d is downcomer cross-sectional area, m^2; A_r is riser cross-sectional area, m^2; u_{Gr} is superficial gas velocity in the riser, $m\ s^{-1}$; and C_s is solids concentration, % (w/v) or g dry weight per 100 ml. Equation (30) was determined using A_d/A_r ratios of 0.25 and 0.44, and $0.026 \leq u_{Gr}$ $(m\ s^{-1}) \leq 0.21$. The solids tested were Solka Floc fibres at concentrations between 1% and 3% in 0.15 M NaCl.

2.3.3 Shear

Mechanical damage in plant-cell suspensions depends on the intensity of shear developed in the reactor. A limited number of correlations for hydrodynamic shear in stirred and airlift reactors is available.

2.3.3.1 Stirred Tanks

Equation (9) is widely accepted as a means for calculating average shear rates in stirred vessels. Variation of shear has been investigated by Metzner and Taylor [54] and Oldshue [85]; maximum shear levels occurring at the impeller tip are significantly higher than those given by Eq. (9).

Wichterle et al. [86] measured maximum shear rates on the front surface of a Rushton 6-blade turbine in non-Newtonian pseudoplastic fluids. For $Re_i > 10$ the following relationship was determined:

$$\frac{\dot{\gamma}_{max}}{N_i} = (1 + 5.3n)^{1/n} \left(\frac{N_i^{2-n} D_i^2 \varrho_L}{K}\right)^{1/(1+n)} \tag{31}$$

where n is the flow behaviour index for power-law fluids and K is the consistency index.

2.3.3.2 Airlift Reactors

At the present time, reliable estimation of average shear levels in airlift reactors is difficult. Correlations reported for pneumatically agitated columns vary considerably, as summarised in Table 4. Of these relationships, the equation by Nishikawa et al. [67] is used almost exclusively; it was developed by analysing heat-transfer coefficients in Newtonian and non-Newtonian liquids in a bubble column at $0.04 \le u_G$ $(m\,s^{-1}) \le 0.1$. The correlation of Henzler [87] was determined by fitting literature data on mass transfer coefficients obtained with non-Newtonian liquids. The broad lack of agreement between the shear-rate correlations of Table 4 causes difficulties in estimation of mass transfer coefficients and mixing times; correlations such as Eqs. (13) and (27) depend on estimation of μ_{app} which, for non-Newtonian fluids, relies on knowledge of $\dot{\gamma}$.

Constant but relatively low shear-stress is produced in air-driven reactors by flow of fluid at the walls of the vessel. Tramper et al. [88] estimated the shear stress

Table 4. Effective shear rate correlations for pneumatically agitated reactors. (Adapted from Schumpe and Deckwer [211])

Correlation	Reactor type	Reference
$\dot{\gamma} = 5000u_G$	bubble column	[67]
$\dot{\gamma} = 1500u_G$	bubble column	[87]
$\dot{\gamma} = 2800u_G$	bubble column	[211]
$\dot{\gamma} \propto u_G/D_c$	bubble column	[82]
$\dot{\gamma} = u_B/D_b$	bubble column	[212]
$\dot{\gamma} = \left(\frac{P/V_L}{K}\right)^{1/(n+1)}$	bubble column	[213]

associated with frictional wall losses in an airlift reactor using the friction-factor equation:

$$\tau_w = \frac{1}{2} \varrho_L u_L^2 k_w \tag{32}$$

where τ_w is shear stress at the wall, Pa; ϱ_L is liquid density, kg m^{-3}; u_L is fluid velocity, m s^{-1}; and k_w is a resistance coefficient. Maximum wall shear-stress was assumed to occur where the direction of fluid flow changes abruptly. In external-loop airlift vessels this happens at the top and bottom of the vessel in the connections between riser and downcomer. In internal-loop reactors energy losses at the top will usually be minimal relative to the bottom because the head-space of internal-loop devices is like an open channel whereas the flow path at the bottom is much more constricted. For maximum shear stress at the vessel wall, the value for k_w in Eq. (32) can be taken to be 1.3 as for a sharp bend.

Much higher levels of shear are associated with gas sparging in air-driven bioreactors than with fluid flow. Shear stress produced in airlift reactors by injection of air bubbles, rising of the bubbles through the liquid, and bubble bursting have been analysed by several groups [32–34, 89]. From calculations [89] and supporting experiments [32], it appears that bursting bubbles create the highest shear stresses in bubble columns. Shear levels associated with bubble break-up have been estimated at 104 N m^{-2} [89] and 200–300 N m^{-2} [34]. Duration of these high shear forces is of the order of milliseconds [34].

2.4 Calculation of Time Constants in Plant-Cell Reactors

Comparison of time constants for different reactor functions can be used to determine which processes are rate-limiting. Time constant or characteristic time is defined as the ratio of capacity to flow; a small time constant represents a fast process.

In the following analysis, time constants are used to assess the suitability of stirred-tank and airlift reactors for culture of plant-cell suspensions. The procedure follows closely the technique of Kossen [90]. Typical geometries for the reactors and other system parameters are assumed, as specified in Table 5. Calculations are performed for an external-loop airlift reactor and for a baffled stirred-tank operated with a flat-blade turbine-impeller and a constant air superficial velocity of 0.05 m s^{-1}. The volume of both reactors is 10 m^3.

The time constants used in this comparison are those for oxygen conversion (t_{rxn}), mass transfer (t_{mt}), and mixing (t_{mx}). Processes of heat transfer and heat production are not considered since they are not expected to cause problems during plant-cell culture. For stirred reactors the time constants are presented as functions of the main operating variable, the stirrer speed N_i. For airlift devices the operating variable is the gas superficial velocity in the riser, u_{Gr}. Since the effectiveness of mixing and mass transfer in plant-cell reactors changes considerably as biomass levels increase, the analysis is performed using conditions for two different times during batch or fed-batch culture. Near the beginning of the culture

Table 5. Parameter values for evaluation of time constants

Parameter		Stirred tank	External-loop airlift reactor
C^*	solubility of oxygen in culture liquid ($kg\,m^{-3}$)	8×10^{-3}	8×10^{-3}
C_s	solids concentration (%)		
	near beginning of growth phase	0.5	0.5
	near end of growth phase	3	3
D_d	downcomer diameter (m)	–	0.78
D_i	stirrer diameter (m)	0.78	–
$\mathscr{D}_L$	diffusivity of oxygen in culture liquid ($m^2\,s^{-1}$)	2.3×10^{-9}	2.3×10^{-9}
D_r	riser diameter (m)	–	1.2
D_T	diameter of stirred tank (m)	2.3	–
g	gravitational acceleration ($m\,s^{-2}$)	9.8	9.8
H_L	height of liquid in stirred tank (m)	2.4	–
H_d	downcomer height (m)	–	6
H_r	riser height (m)	–	6
u_G	superficial gas velocity ($m\,s^{-1}$)	0.05	varies
V_L	volume of liquid in the vessel (m^3)	10	10
X	cell concentration (kg dry weight m^{-3})		
	near beginning of growth phase	5	5
	near end of growth phase	30	30
ϱ_L	density of culture liquid ($kg\,m^{-3}$)	1015	1015
σ	surface tension ($kg\,s^{-2}$)	0.04	0.04

the cell concentration is relatively low at 5 $kg\,m^{-3}$ dry weight; the situation is then re-assessed for conditions at the end of growth when 30 $kg\,m^{-3}$ cells are present.

The equations which will be used to calculate the time constants have been presented in the preceding sections. As already noted, reactor-design equations for non-Newtonian liquids are not as well defined as for low-viscosity Newtonian fluids, and correlations for mass-transfer coefficients are generally limited due to the sensitivity of this parameter to medium composition and method of measurement. Application of equations other than those used below may give different results; however, the overall picture does not change dramatically. None of the available correlations for mixing and mass transfer were developed with plant-cell cultures; virtually all bioreactor studies of the effect of solids and high viscosity have used cellulose-fibre suspensions or homogeneous solutions such as carboxymethyl cellulose (CMC) to simulate cell slurries. The deficiencies of homogeneous fluids as a substitute for cell suspensions have been discussed by Allen and Robinson [68] and, although Solka Floc fibre suspensions show a strong resemblance to mycelial cultures such as *Aspergillus* and *Penicillium* [91], they are not a close replica of plant-cell suspensions. This rheological dissimilarity represents a major limitation of the following theoretical study. However, until further work is completed with plant cells, correlations derived from simulation studies are the only ones available for modelling plant-cell reactors.

2.4.1 Oxygen Consumption

During growth, the time constant for oxygen consumption t_{rxn} can be expressed as:

$$t_{rxn} = \frac{C^*}{r_{O_2}} \tag{33}$$

where C^* is the equilibrium concentration of oxygen in the culture liquid, kg O_2 m^{-3}, and r_{O_2} is the volumetric oxygen consumption rate, kg O_2 m^{-3} s^{-1}. The value for r_{O_2} will depend on the biomass concentration X, kg m^{-3}, and the observed specific oxygen consumption rate q_{O_2}, kg O_2 kg^{-1} s^{-1}. Note that q_{O_2} is an observed parameter and so incorporates a maintenance contribution.

In the literature, oxygen consumption by plant cells in batch culture does not show a consistent pattern. Data reported by Kobayashi et al. [74] for *Thalictrum minus* suspensions and by Hong et al. [38] for strawberry suspension cultures show that specific oxygen-uptake rates increase during growth to reach a maximum near the onset of stationary phase. In contrast, Townsley et al. [92] and Drapeau et al. [93] demonstrated with *Triptergium wildordii, Catharanthus roseus* and *Dioscorea deltoidea* cultures that specific rate of oxygen consumption declines progressively throughout batch culture. Quantitative results from these previous studies are summarised in Table 6.

In the present analysis q_{O_2} will be considered constant over the growth period and equal to an average value of 10^{-6} kg O_2 kg dry weight^{-1} s^{-1}. Combining these figures with the biomass concentrations, at the beginning of growth r_{O_2} is 5×10^{-6} kg O_2 m^{-3} s^{-1} and t_{rxn} is 1600 s; at the end of growth r_{O_2} is 3×10^{-5} kg O_2 m^{-3} s^{-1} and t_{rxn} is 270 s. The calculated O_2 uptake rate at the end of growth is greater than the maximum rate of 1.6×10^{-5} kg O_2 m^{-3} s^{-1} reported by Scragg and Fowler [94] and Drapeau et al. [93]; this is probably due to the higher cell

Table 6. Literature values for observed specific oxygen uptake rate q_{O_2} in suspended plant-cell cultures

Plant	q_{O_2} kg O_2 (kg dry weight)$^{-1}$ s^{-1} $\times 10^6$	Reference
Strawberry	1.1–1.5 (increasing during growth)	[38]
Thalictrum minus	0.18–4.3 (increasing during growth)	[74]
	q_{O_2} kg O_2 (kg fresh weight)$^{-1}$ s^{-1} $\times 10^6$	
Catharanthus roseus	0.21–0.03 (decreasing during growth)	[93]
Dioscorea deltoidea	0.51–0.08 (decreasing during growth)	[93]

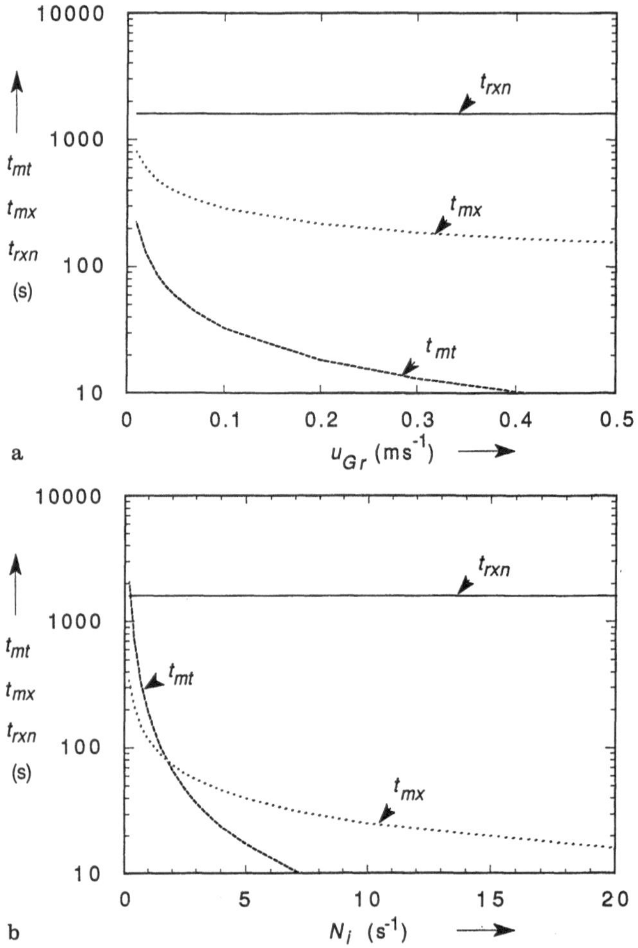

Fig. 5a, b. Mixing, mass transfer and oxygen consumption in bioreactors at a suspended plant-cell concentration $C_s = 5\ \mathrm{kg\ m^{-3}}$ dry weight **a)** external-loop airlift reactor; **b)** stirred-tank reactor

density. The calculated time constants for oxygen uptake are plotted in Figs. 5 and 6. For comparison, time constants for microbial oxygen consumption are of the order 9–45 s [90].

2.4.2 Mass Transfer

The expression for the mass-transfer time constant is:

$$t_{mt} = \frac{1}{k_L a_L} \tag{34}$$

where $k_L a_L$ is the mass transfer coefficient based on gas-liquid interfacial area per unit volume of liquid.

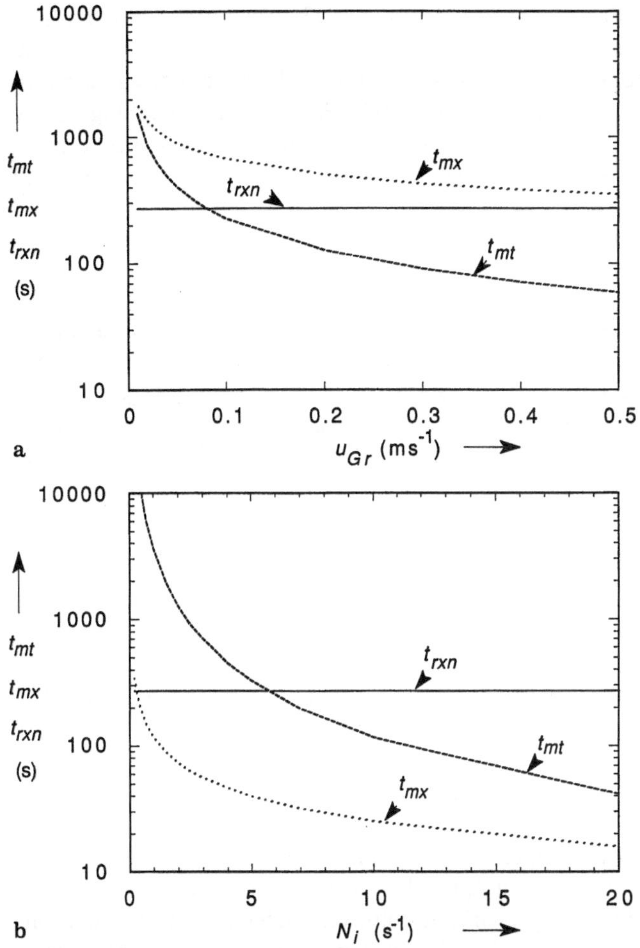

Fig. 6a, b. Mixing, mass transfer and oxygen consumption in bioreactors at a suspended plant-cell concentration $C_s = 30\ kg\ m^{-3}$ dry weight. **a)** external-loop airlift reactor; **b)** stirred-tank reactor

2.4.2.1 Airlift Reactors

For airlift reactors the value of $k_L a_L$ can be calculated from Eqs. (27)–(29) or from Eq. (30). In plant-cell cultures, high viscosity is due mainly to the solids content, so both Eqs. (27) and (30) apply.

$k_L a_L$ values have been calculated from Eq. (30) with $C_s = 0.5\%$ and $C_s = 3\%$. Corresponding values for the mass-transfer time constant (t_{mt}) for gas velocities between 0.01 m s^{-1} and 0.5 m s^{-1} are plotted in Figs. 5a and 6a. Final $k_L a_L$ values at the end of growth are about 14% those near the beginning.

If $k_L a_L$ results calculated from Eqs. (27)–(29) and (30) are equivalent, μ_{app} values can be determined to check the flow conditions. Depending on the gas flow rate and therefore the shear rate in the reactor, apparent viscosities at the beginning of culture range from 36 mPa s at $u_{Gr} = 0.01$ m s^{-1} to 13 mPa s at $u_{Gr} = 0.5$ m s^{-1}. At the end of the growth period, the corresponding μ_{app} values are 320 Pa s and 110 mPa s. The fluid consistency index K and flow behaviour index n can also be calculated from Eqs. (3) and (15): K is 0.10 Pa s^n at the beginning of the culture and 0.99 Pa s^n at the end; n = 0.71 irrespective of cell concentration. These calculated values for apparent viscosity, K and n are realistic for plant-cell suspensions.

2.4.2.2 Stirred Tanks

Since data on the viscoelastic properties of plant-cell suspensions are not available Eq. (23) cannot be easily applied for calculation of k_La_L. Instead, with the above values for K and n, Eqs. (23)–(25) are used to evaluate k_La_L for oxygen transfer in stirred vessels. Results for the mass-transfer time constants are plotted in Figs. 5b and 6b. Depending on the stirrer speed between 0.2 s^{-1} and 20 s^{-1}, μ_{app} values are 88 mPa s to 23 mPa s at the beginning of culture and 850 mPa s to 220 mPa s at the end.

2.4.3 Mixing

2.4.3.1 Airlift Reactors

Mixing times for external-loop airlift reactors can be calculated from Eqs. (13), (16), (18), (19) and (8). The results are plotted in Figs. 5a and 6a.

2.4.3.2 Stirred Tanks

Mixing times for stirred-tank reactors are obtained from the Norwood-Metzner correlation of Eq. (10). The value of f_m is taken to be 20 as determined by Wang and Fewkes [58] with *Streptomyces niveus* cultures. Under turbulent conditions the mixing time in stirred reactors is largely independent of viscosity; consequently t_{mx} values are the same at the end of the plant-cell culture as at the beginning. The results are plotted in Figs. 5b and 6b.

2.4.4 Shear

2.4.4.1 Airlift Reactors

Shear stress in air-driven reactors is currently the subject of extensive investigation. As discussed in Sect. 2.3.3, recent studies have suggested that highest shear is associated with bubble disengagement. However, the duration of these forces is extremely short, of the order of milliseconds. Although these conditions might be sufficient to cause damage to fragile animal cells without cell walls, it is unclear whether the same mechanism is responsible for plant-cell damage. In this analysis, it is assumed that the less intense but more prolonged shear-forces associated with fluid flow determine whether shear damage is suffered by suspended plant cells in airlift reactors.

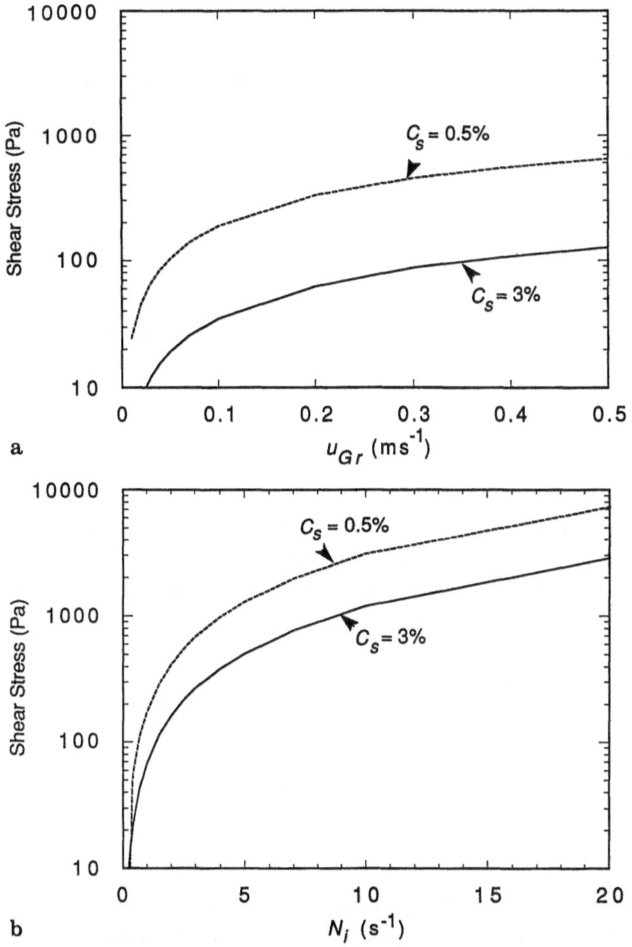

Fig. 7a, b. Liquid shear stress produced in bioreactors containing suspended plant-cell concentrations $C_s = 5$ kg m^{-3} dry weight (0.5%) and $C_s = 30$ kg m^{-3} (3%). **a)** airlift reactor; **b)** stirred-tank reactor

Fluid shear-stress is maximum in airlift reactors in the top and bottom sections between riser and downcomer where flow directions change abruptly. The magnitude of this stress is calculated from Eq. (32). Results for the beginning and end of plant-cell culture are plotted in Fig. 7a.

2.4.4.2 Stirred Tanks

In stirred reactors, maximum shear levels associated with the action of the impeller are used to determine whether plant cells are likely to be disrupted by hydrodynamic forces. Shear stress at the impeller tip is calculated from Eq. (31) for a 6-flat-blade Rushton-type agitator. Results for the beginning and end of the culture are shown in Fig. 7b.

2.5 Comparison of Reactor Performance

The general trend of calculated results shown in Figs. 5 and 6 gives a realistic picture of reactor performance.

The mixing and mass-transfer situation in stirred and airlift reactors at a plant-cell concentration of 5 kg m^{-3} dry weight is represented in Fig. 5. As reported in the literature, performance of the airlift reactor at this biomass density is adequate; both mass-transfer and mixing have shorter time constants than oxygen consumption. In the stirred reactor conditions are similar; the results show that only a low stirrer speed is required for adequate oxygen mass transfer. Because the correlations for stirred-tank k_La_L have limited applicability at low Re_i below $N_i = 0.3\ s^{-1}$, the actual stirrer speed required for equal values of t_{rxn} and t_{mt} should not be read directly from Fig. 5b.

By the end of the culture when cell concentration is 30 kg m^{-3}, the situation is somewhat different. Figure 6a shows how mixing becomes limiting in the airlift reactor at high cell densities; at gas velocities below 0.5 m s^{-1} mixing times are larger than the oxygen-consumption time so that oxygen gradients will develop. This is consistent with experimental findings that mixing problems develop in airlift vessels at biomass densities above about 25 kg m^{-3} dry weight [36, 37, 44]. Mass transfer requirements are satisfied at gas flow rates above about 0.08 m s^{-1}. From Fig. 6b, mixing in the stirred vessel is very effective even at low stirrer speeds; however the mass-transfer time constant is greater than the reaction time constant until $N_i = 6\ s^{-1}$. At $N_i > 6\ s^{-1}$ adequate oxygen transfer and mixing are provided.

These results for mixing and mass transfer must be considered in conjunction with predicted shear levels in the reactors (Fig. 7). The maximum shear stress tolerated by the cells will set upper limits on N_i and u_{Gr}. Unfortunately, as discussed above, tolerance of plants cells to shear is unclear at present. However, as an example, assume the critical shear stress for growth of plant cells is 100 Pa, 25–100 times higher than that reported for insect cells [27]. From Fig. 7a, operation of airlift reactors at the beginning of the culture must therefore be limited to gas velocities less than 0.05 m s^{-1}; from Fig. 5a this does not adversely affect mixing or mass transfer. As cell density increases, higher gas velocities up to about 0.3 m s^{-1} can be used since the shear stress developed in the liquid decreases with increasing viscosity. In the stirred reactor (Fig. 7b), N_i must be less than about 0.5 s^{-1} at the beginning and less than 1.5 s^{-1} towards the end of growth to avoid shear damage to the culture. Although the reactor will function reasonably well at low stirring speeds at the beginning of the culture (Fig. 5b), the requirement for low N_i at the end of growth conflicts directly with the necessity for $N_i > 6\ s^{-1}$ for proper mixing and mass transfer (Fig. 6b). At the assumed level of shear sensitivity, stirred reactors cannot provide non-destructive conditions for plant-cell culture.

The above analysis is subject to limitations in addition to those already discussed. If plant-cell aggregates are present, transport of oxygen depends not only on k_La_L but also on diffusion through the cell pellet. Accordingly, adequate gas-liquid mass transfer is provided when $t_{mt} < t_{rxn}$; however oxygen deficiencies may still occur in the interior of cell aggregates. In addition, the analysis focuses solely on requirements for growth; t_{rxn} values were calculated using specific oxygen-uptake rates during active growth. In interpretation of Figs. 5 and 6 it was assumed that if time constants for mixing and mass transfer are smaller than for oxygen consumption then reactor conditions are favourable. In the case of secondary production however, high rates of mass transfer may not be desirable as oxygen deficiency is known to stimulate secondary-metabolite synthesis in some cases. In theory, a similar set of calculations could be carried out to test whether the reactors are able to provide mixing, mass-transfer and shear conditions for optimal productivity. This is difficult at the present time because the conditions required for secondary synthesis are generally not known.

2.6 Improvement of Reactors

In the above analysis, the various functions of bioreactors: mixing, mass transfer and provision of shear, were considered separately to determine which limits growth under different culture conditions. Once the limiting process is identified, measures can be taken to improve it.

2.6.1 Airlift Reactors

At high cell density, performance of airlift reactors is limited by mixing. Although adequate mixing is provided at high gas velocities, increasing the gas flow rate may not be a practical solution. Foaming and foam separation of cells is a problem in plant-cell suspensions; to avoid formation of cell meringue at the top of the vessel high gas flow should be avoided. Antifoams non-toxic to plant cells can be used to relieve this problem [95, 96]; however significant reduction in k_L can occur after addition of antifoam [97]. Scragg, Fowler and colleagues have identified another problem with high gas throughput in plant-cell cultures: volatile media components are stripped from the liquid at high ventilation rates to the detriment of cell growth [39, 40].

Mixing in airlift reactors must be increased by means other than raising the gas flow rate. In some cases, attention to geometry: A_d/A_r ratio, aspect ratio, draft-tube clearances and flow-path smoothing (Fig. 4), will give some improvements to liquid circulation, as will choice of external-loop rather than internal-loop designs. Multiple gas-injection points in the reactor may also be beneficial. Chisti [50] suggests that separate sparging of gas into the downcomer could be useful; study of this mode of operation is needed to identify instability conditions and monitor the effects on mass transfer and liquid flow. Adding an impeller to the airlift design should also improve liquid circulation as long as shear levels are minimised. An axial-flow propeller installed inside the draft tube of an airlift reactor has been tested with plant-cell suspensions; for *Beta vulgaris* a stirrer speed of 28 rpm caused sufficient shear damage to reduce biomass and product yields [9].

Because increased cell content is responsible for the decline of mixing efficiency in airlift reactors, another strategy would be to reduce cell concentration and culture viscosity. Dilution of plant-cell suspensions by small volumes of water or fresh medium can significantly reduce viscosity. According to the relationship between viscosity and plant-cell concentration reported by Tanaka [10], a 10% dilution will lower apparent viscosity by about 50% as reported previously for suspensions of filamentous mycelia [55]. Semi-continuous processes, where a volume of culture is withdrawn periodically from the reactor and nutrient medium added, would be advantageous for plant cells as this allows a far greater reduction in viscosity than continuous feeding. Draw/fill methods for airlift reactors have been tested by Scragg et al. [42] with *Catharanthus roseus.*

Although improvements are possible, if plant-cell densities substantially greater than 30 kg m^{-3} dry weight are required for economic feasibility, it is likely that even the optimal airlift reactor will not be able to provide adequate mixing and mass transfer conditions.

2.6.2 Stirred Tanks

In stirred vessels, sufficient mass transfer and mixing occur at stirrer speeds above about 6 s^{-1}. Although this is a relatively low speed for most microbial processes, for shear-sensitive cells the shear stress imposed by a Rushton-type impeller under these conditions would be destructive. Improvement of stirred reactors for plant-cell culture can therefore be distilled down to reduction of shear levels.

Some studies of low-shear impellers have been carried out with plant-cell suspensions. Tanaka [36] tested a large modified paddle-type impeller in a bioreactor with four small baffles; this stirrer was operated at 60 rpm and, after sucrose feeding, a cell density of 30 kg m^{-3} was achieved.

Helical screw impellers (Fig. 8) have been studied extensively [51, 98] for mixing viscous materials; mixing is accomplished without high liquid velocity streams. The helical screw usually functions by carrying liquid from the vessel bottom and discharging it at the liquid surface; alternatively, the screw may be operated in reverse to pull liquid to the bottom of the vessel. Baffles are generally required in screw-agitated reactors to promote turbulence and avoid dead zones near the vessel wall. Helical agitators have been tested with plant-cell suspensions [37, 99]. A helical stirrer operated at 100 rpm produced high levels of rosmarinic acid from *Coleus blumei* cells without shear damage and with k_La_L values between 1.5×10^{-3}

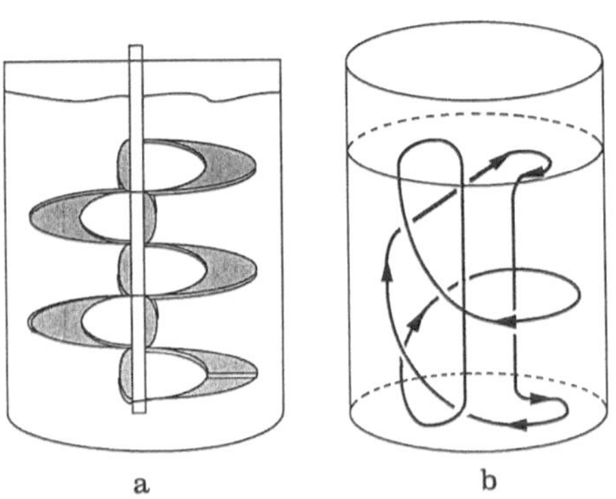

a b

Fig. 8a. Typical configuration of helical screw impeller; **b)** Typical flow pattern produced by a helical impeller. (Adapted from Bourne and Butler [98])

and 6.0×10^{-3} s^{-1}. In the same study an anchor impeller operated at a maximum speed of 40 rpm produced significantly lower product levels.

Hong et al. [38] installed 25 mm-wide teflon ribbons between the blades of two turbine impellers in a 7-litre stirred bioreactor in an attempt to reduce shear and improve mixing. The modified agitator was tested for culture of strawberry cells; growth did not occur and the viability of the cells dropped from 80% to less than 25% in 4 days. Better results were obtained by Hooker et al. [19] with a large flat-blade impeller with and without sail-cloth blades. Higher maximum growth rates were obtained compared with a smaller flat-blade turbine. Power consumption by large sail-cloth impellers is considerably greater than with Rushton turbines [51].

2.7 Alternative Reactor Designs

As well as modifying airlift and stirred reactors to improve plant-cell culture, alternative reactor configurations have been sought. The most promising of these is the rotating-drum reactor investigated by Tanaka et al. [100]. This design was found to be superior to mechanically agitated vessels; oyxygen transfer was sufficient to achieve a biomass density of 20 kg m^{-3} dry weight under conditions of high viscosity and low hydrodynamic stress. Tanaka and co-workers determined a correlation for $k_L a_L$ in the drum as a function of rotational speed, number of baffles and baffle height, tank diameter and fraction of the tank volume filled with liquid. The decrease in $k_L a_L$ due to apparent viscosity was smaller in the rotating drum than in mechanically agitated reactors.

3 Immobilised Plant Cells

Early reports by Brodelius et al. [101] showed that phytochemical synthesis is enhanced if plant cells are first entrapped in alginate gel. The feasibility of immobilised plant cells has since been demonstrated for single- and multiple-step biotransformations [102–104] and for de novo synthesis of secondary metabolites [105, 106].

Aspects of immobilised plant-cells have been extensively reviewed in recent years. Discussion of potential technical and physiological advantages can be found in articles by Shuler et al. [107], Lindsey and Yeoman [108] and Robins et al. [109]. Established techniques for immobilising plant cells by entrapment or surface-attachment are also described [106, 110–114]. Of the natural gels such as alginate, agarose, agar, gelatin and carrageenan commonly used for cell entrapment, alginate appears to be most beneficial for plant cells. Studies comparing the effects of these materials have shown that better secondary product formation and excretion occur when alginate is used [115, 116]. This effect has not yet been explained; Hulst et al. [117] determined that the oxygen-transfer characteristics of alginate were not significantly different from other gels. Ayabe et al. [118] report that production of retrochalcone by freely suspended *Glycyrrhiza echinata* cells is increased if calcium alginate beads are added to the medium; Prenosil et al. [119] have postulated that alginate, being similar to the polycarboxylic acids in cell walls, mediates cell-cell interactions in the presence of calcium ions to stimulate secondary synthesis. Enhanced secondary product synthesis has also been reported for plant cells immobilised in reticulated polyurethane foam [106, 120, 121].

For most efficient operation, the product of interest should be excreted from immobilised cells into the medium so that the biomass does not have to be periodically destroyed. There is evidence that immobilisation enhances release of secondary products from plant cells [101, 122]. Alternatively, release of compounds can be induced by permeabilising agents such as DMSO [123], although limited success has been achieved using this approach. Once in the medium, products can be extracted in situ using adsorptive resins such as Amberlite XAD-7 [124] or immiscible solvents such as vegetable oil [125] or n-hexadecane [126]. Continuous product extraction encourages further synthesis and minimises degradative metabolism.

Most techniques for immobilising plant cells produce an environment in which there are substrate, oxygen, hormone and other concentration gradients. This applies even to methods involving surface attachment; once a monolayer of plant cells is laid down, other cells readily adhere to it to form a thick biofilm. Designing large-scale immobilised-cell processes requires knowledge of growth and production kinetics. To determine whether mass transfer limits plant-cell activity, analysis of diffusion processes must be carried out.

3.1 Mass Transfer Considerations

Equations for diffusion and reaction are applied to immobilised cells for prediction of effective reaction rates. Intraparticle diffusion is the main resistance to mass transfer in entrapped-cell systems, although external boundary-layers surrounding the particle can also contribute. Despite the relatively slow metabolic rate of plant cells, effects of diffusion on activity of immobilised plant cells cannot generally be neglected.

As well as reducing the overall reaction rate, diffusional limitations mask intrinsic kinetics. Care must be exercised when comparing the performance of different immobilised-cell systems; apparently higher rates of growth or product synthesis in one system may reflect an absence of diffusional effects rather than any intrinsic metabolic response.

3.1.1 External Mass-Transfer Resistance

External mass-transfer resistance is largely under the control of reactor hydrodynamics; increasing turbulence and mixing reduces the thickness of the liquid boundary-layer and improves supply of nutrients to the particle surface. External mass transfer is most likely to contribute to overall diffusion resistance when reactors are not well mixed, for example in packed beds.

The significance of external diffusion in reactor systems can be determined. In most cases, if external mass transfer is affecting overall reaction rate, improved mixing increases the observed reaction velocity [127]. On the other hand, if the rate of external diffusion is much more rapid than the rate of reaction, the observed rate should be largely independent of mixing intensity.

3.1.2 Internal Mass-Transfer Resistance

Intraparticle diffusion in immobilised plant-cell systems has been analysed by Hulst et al. [117, 128], Mavituna et al. [129], Furusaki et al. [130] and Pras et al. [131]. Whether or not internal mass-transfer of substrate affects reaction rate depends on several factors:

a) specific cellular demand for substrate, mol $cell^{-1}$ s^{-1};
b) reaction kinetics, i.e., the relationship between reaction rate and substrate concentration;
c) biomass loading in the matrix (or biofilm), cells m^{-3};
d) particle diameter (or film thickness), m;
e) effective diffusivity of substrate molecules through the immobilisation matrix (or biofilm), m^2 s^{-1}; and
f) concentration of substrate in the medium at the exterior surface of the particle (or biofilm), mol cm^{-3}.

Since these parameters vary considerably from system to system, results from one analysis cannot be readily applied to other immobilised-cell situations.

The molecular species most likely to cause diffusional limitations in immobilised plant-cells must be determined before the analysis can proceed. Mavituna et al. [129] analysed diffusion of glucose into reticulated foam containing *Capsicum frutescens* cells and determined that mass transfer of this substrate was not an important problem. Furusaki et al. [130] measured rates of sucrose, oxygen and codeinone consumption by immobilised *Papaver somniferum* cells; while effectiveness factors for these substrates were close to unity, calculations showed that the concentration of oxygen decreased considerably within the particles. Pras et al. [131] concluded that limitations in oxygen transport were most likely after measuring the effective diffusion coefficients and cellular demands for oxygen and L-DOPA precursors by *Mucuna pruriens* entrapped in alginate gel.

In the case of oxygen as limiting substrate, for a spherical particle containing plant cells, if the following assumptions can be made:

a) the cells are identical and their distribution in the particle is uniform;
b) oxygen transport in the particle occurs only by diffusion;
c) oxygen gradients occur only in the radial direction;
d) the effective diffusion coefficient for oxygen is independent of oxygen concentration;
e) respiration can be described using a Michaelis-Menten kinetic model; and
f) the number of plant cells is constant;

mass transfer and consumption of oxygen at steady state is described by the mass-balance equation:

$$\mathscr{D}_e\left(\frac{1}{r^2}\frac{d}{dr}\left(r^2\frac{dC}{dr}\right)\right) = \frac{v_{max}C}{K_m + C} \tag{35}$$

with the usual boundary conditions:

$dC/dr = 0$ at $r = 0$ (centre of the particle)

$C = C_e$ at $r = R$ (surface of the particle).

In Eq. (35), r is radial distance measured from the centre of the particle, m; C is oxygen concentration within the particle, mol m^{-3}; $\mathscr{D}_e$ is the effective diffusion coefficient for oxygen in the particle containing cells, $m^2\ s^{-1}$; v_{max} is the maximum reaction velocity, mol $m^{-3}\ s^{-1}$; K_m is the Michaelis-Menten constant, mol m^{-3}; C_e is the oxygen concentration at the surface of the particle, mol m^{-3} and R is the radius of the particle, m. Equation (35) can be solved only by numerical integration to give the oxygen profile in the particle, C as a function of r. A similar though simpler equation is obtained for a biofilm which can be assumed infinitely long and in which diffusion of oxygen occurs only in one direction.

Hulst et al. [128] calculated using Eq. (35) the particle size at which the oxygen concentration in the centre is zero. The analysis was performed for cell aggregates containing 100% *Tagetes patula* cells with a dry-weight cell-density of 10 kg m^{-3}, and intrinsic Michaelis-Menten constants $v_{max} = 4 \times 10^{-3}$ mol $m^{-3}\ s^{-1}$ and $K_m = 3.5 \times 10^{-2}$ mol m^{-3}. $\mathscr{D}_e = 1.9 \times 10^{-9}\ m^{-2}\ s^{-1}$ and $C_e = 0.25$ mol m^{-3}. Under these conditions, the critical particle diameter for depletion of oxygen at the centre was 3 mm.

Some restrictions apply to use of the diffusion-reaction analysis represented by Eq. (35). Equation (35) is a valid description of mass transfer and reaction under conditions of constant biomass concentration in the immobilised-cell particle. However, for systems in which growth occurs, the value of v_{max} varies according to the quantity of cells present. When cell concentration increases in alginate beads, reticulated foam and membrane units, oxygen concentration gradients and diffusional limitations will become more severe with time. In addition, use of Eq. (35) assumes knowledge of the intrinsic kinetic parameters for immobilised cells, v_{max} and K_m. Experimental determination of these values is not straightforward [128, 131] and use of suspended-cell parameters may not be valid.

Although observed rates of oxygen consumption are often lower than in suspension due to diffusion limitations, immobilised plant-cells still perform secondary synthesis. The exact response to oxygen starvation depends on the particular cell system and product of interest. Mavituna et al. [125] report that capsaicin synthesis by foam-entrapped *Capsicum frutescens* is enhanced when the bulk-liquid dissolved-oxygen concentration is almost zero; if aeration is restored capsaicin in the medium disappears. Pras et al. [131] showed that production of

L-DOPA by immobilised *Mucuna pruriens* continues in the presence of significant oxygen limitation. Kargi et al. [132] measured oxygen profiles in biofilms of *Catharanthus roseus* and found that, although the oxygen concentration was close to zero approx. 2 mm from the top surface, indole-alkaloid formation was stimulated and highest alkaloid levels were found in biofilms of about 4 mm thickness. It appears that although oxygen supply in immobilised cells is often restricted due to diffusion limitations, oxygen concentration does not necessarily control productivity of the cells. Alternatively, as suggested by Lindsey and Yeoman [108], secondary metabolism and cell differentiation may be enhanced in response to strong oxygen concentration gradients.

3.2 Self-Immobilised Plant Cells

The ability of plant cells to spontaneously aggregate into macroscopic clumps is well known and often causes operating problems during large-scale suspension culture. However tendency to clump can in some cases eliminate the need for artificial immobilisation supports and could be exploited for large-scale phytochemical synthesis. The success of this approach relies on formation of compact propagative aggregates of regular shape without large quantities of dispersed cells also being present. Self-immobilisation of plant cells has not received as much attention as gel- or foam-entrapment; the range of species which can be self-immobilised and the long-term feasibility of this type of plant biocatalyst is unclear. However, there is an increasing number of reports describing formation and application of aggregated cells, including *Populus alba* [133], *Jasminum* species [134], *Coffea arabica* [119], *Tagetes patula* [128], *Cinchona ledgeriana* [135], and *Solanum aviculare* [136].

General procedures for self-immobilisation of plant cells are described by Fuller and Bartlett [137]. Enhanced alkaloid synthesis compared with suspended cultures has been found [135, 136]; these authors also report organisation of cells within the aggregates and evidence of tracheid differentiation. Nutrient starvation in larger aggregates causes cell lysis so that the centres are hollow. Hulst et al. [128] correlated thiophene production with aggregate diameter; mean thiophene levels increased with size up to 12 mm.

3.3 Immobilised-Cell Reactors

Bioreactor design for scale-up of immobilised plant cells is at an early stage; suitability of different configurations for immobilised-cell culture is still being tested. Application of established reactor designs such as stirred tanks, packed and fluidised beds and airlift vessels has been described in previous review papers [113, 114, 138]. Self-immobilised plant cells have requirements similar to artificially-immobilised cells in terms of reactor design, and have been cultured successfully in fluidised-bed [119] and conical bubble-column [136] reactors.

Important criteria for bioreactor design for immobilised plant-cells include:

a) good bulk-fluid mixing to minimise external mass-transfer resistances;
b) adequate exposure to air or aerated medium; and
c) minimal shear damage.

Mode of aeration in immobilised plant-cell reactors appears to have an important influence on cell activity. In early work, Veliky and Jones [103] found that production of 5β-hydroxygitoxigenin in a packed-bed reactor was improved if air were sparged directly into the column; if the medium were aerated away from the reactor and recirculated through the bed, bioconversion was lower and cells quickly lost viability. Kobayashi et al. [139, 140] devised a modified packed-bed reactor for alginate-entrapped *Thalictrum minus* cells to provide direct contact with air. Operation involved periodic exposure of the beads to liquid medium and then to air; optimal cycle conditions for berberine production were 2 min exposure to medium then 30 s air supply. Direct contact with air appeared crucial for synthesis of berberine; when beads were continuously soaked in medium saturated with oxygen, production was almost completely suppressed.

As well as the standard types of bioreactor, novel designs have also been tested. Special reactors have been devised for surface-attached cells and biofilms; surface-immobilisation of plant cells has the aim of reducing diffusion limitations while maintaining high levels of cell-cell contact. Archambault et al. [141, 142] attached *Catharanthus roseus* cells to non-woven, short-fibre polyester; the fabric was wound into a square-spiral configuration on stainless-steel supports and placed into airlift and mechanically-agitated vessels. A similar arrangement was also tested with *Papaver somniferum* cells for sanguinarine production [143]. Plant cells were immobilised in situ; for *P. somniferum*, suspended cells became attached to the matrix after 10–15 min exposure to the fabric and the entire surface was covered within 6d. Kargi et al. [132] describe a horizontal-biofilm reactor containing plant cells growing on nutrient agar over which liquid medium is continuously recirculated. The optimum biofilm thickness for maximum alkaloid production was determined using this device.

The problem of intracellular storage of products in immobilised plant-cell systems has been addressed by Tramper et al. [144]. A new type of density-difference bioreactor called a liquid-impelled loop reactor has been designed; organic solvent immiscible in water is introduced into the vessel to cause circulation and mixing in much the same way as air does in airlift reactors. The solvent can be either lighter or heavier than water; in either case the product of interest should partition into the organic phase in preference to the aqueous phase. This has several advantages: in situ extraction of product can be achieved for easier downstream processing, product excretion is enhanced as concentrations in the aqueous phase are reduced, and the chances of product inhibition or further metabolism are lower if direct contact between cells and metabolite is removed. Organic solvents can, however, have a negative effect on cell activity. Final selection of organic solvent must also consider the density difference between the solvent and aqueous phase; a high density difference facilitates mixing in the reactor and phase separation. The liquid-impelled loop reactor has been tested with free and immobilised cells of *Rubia tinctorum* and *Morinda citrifolia*, with hexadecane as solvent [145].

3.4 Effect of Immobilisation on Plant-Cell Metabolism

Interest in immobilising plant cells is motivated to a large degree by reports of enhanced secondary production. There is an expectation that increased cell-cell contact and chemical and electrochemical gradients will mimic conditions in the whole plant to the benefit of secondary synthesis [108]. While enhanced secondary production has been observed in some cases, the actual reasons for this response have not yet been elucidated. Any metabolic changes caused by immobilisation seem to be reversible; upon release of plant cells from calciumalginate gel it has been found that production rapidly returns to the same level as in suspended cells [146].

Little work has been carried out to determine the properties of immobilised plant cells beyond analysis of growth, sugar and oxygen consumption, and alkaloid synthesis. Brodelius [147] compared phosphate uptake and metabolism by free and immobilised *Catharanthus roseus* cells using ^{31}P-NMR; however, no significant differences were found. Mass-transfer limitations in most immobilised plant-cell systems make it difficult to draw conclusions about the effects of immobilisation on cell function. Kargi et al. [132] measured RNA and protein levels in a biofilm of *Catharanthus roseus* attached to agar; cells in the lower layers of the biofilm contained less RNA than those at the top which had better access to nutrients and oxygen. The interior cells were also slower-growing and contained less protein than top-layer cells. Linear growth kinetics and lower biomass yields measured by Archambault et al. [141] for immobilised *Catharanthus roseus* are also most likely due to diffusional limitations; similar responses to oxygen-limited conditions were described previously by Kato et al. [71] and Pareilleux and Vinas [72] for suspended cells. There is evidence, however, that plant-cell immobilisation changes the proportions of secondary products synthesised: Vanek and Macek [148] report that relative yields of three biotransformation products changed depending on whether *Solanum aviculare* cells were freely suspended or immobilised in any of five different support matrices. Increased tendency to secrete non-polar products and synthesis of compounds not detected in suspended-cell culture have been observed for immobilised *Tagetes minuta* [116].

The effects of immobilisation on cell-cycle function and kinetics have not been determined for plant cells. Significant changes in growth patterns and DNA content following surface attachment of microorganisms have been reported [149] but it is not known whether immobilisation of plant cells interferes with cell-cycle operation in a similar way. Ketel et al. [116] have suggested that the rigidity of calcium alginate imposes physical stress on plant cells which stimulates production and excretion of secondary metabolites. It is also possible that spatial restrictions due to the surrounding solid matrix affect cell expansion and division, and limit growth.

The ability of plant cells to differentiate in response to concentration gradients increases the difficulty of separating nutrient-limitation effects from intrinsic metabolic changes caused by contact with the support matrix.

4 Differentiated Plant Tissue

A strong correlation between secondary-metabolite production and morphological differentiation has been shown repeatedly in plant tissue-culture studies; secondary synthesis is often poor in callus and suspensions but recovers immediately after organogenesis [150, 151]. Organogenesis also alleviates problems of instability in de-differentiated plant cells [152, 153]. Large-scale culture of roots is an attractive proposition for root-derived products; roots are able to extend and grow indefinitely as a result of cell reproduction at the tip meristem. Several studies have demonstrated that product levels in roots cultured in vitro are at least as high as those found in the whole plant [154–156]. Embryos are another potential route for secondary-metabolite manufacture; large-scale culture of somatic embryos has additional applications for plant propagation and production of artificial seed. The ability of a large number of plants to form multiple shoots from axillary and shoot-tip meristems could also be used for phytochemical production and for large-scale propagation of elite genotypes. Exploitation of these forms of organised plant tissue on a commercial scale requires design of appropriate reactors. This area is developing rapidly.

4.1 Hairy Roots

Although in vitro culture of excised plant roots was first demonstrated as early as 1934 [157], slow growth of most roots in liquid medium and the requirement for exogenous growth factors are major drawbacks for commercial application. Growth rates can be increased significantly by genetic transformation with *Agrobacterium rhizogenes* bacteria to form 'hairy roots'; these can then be cultured axenically without external hormones. Hairy-root cultures have been tested for production of a broad range of secondary products; members of the Solanaceae family known to synthesise alkaloids in their roots have been studied extensively. Review articles are available discussing the plant host-range for *Agrobacterium rhizogenes*, procedures for infection, methods of hairy-root culture, and the range of compounds produced in vitro [158–161].

In an effort to identify appropriate reactor configurations, several studies have been carried out with hairy roots in batch [162, 163] and continuous-flow air-sparged reactors [164, 165]. Final biomass densities of 10–11 kg m^{-3} dry weight have been obtained [162, 164]. When continuous liquid-flow operation is used, this figure can be increased to about 40 kg m^{-3} dry weight [165]. An important problem with reactor-culture of hairy roots is poor distribution of biomass in the vessels. Growth of hairy roots, whether in shake-flasks or in reactors, usually results in formation of a dense root ball [163, 166]. Nutrient and oxygen limitations in the centre of the ball reduce cell activity and can lead to necrosis. Poor distribution of roots also means that the total volume of the reactor is not used efficiently. To overcome these problems, roots must be more evenly spread and mass transfer improved. Use of conventional stirred reactors is generally not an option for freely-suspended hairy roots; root tissue is easily damaged by mechanical forces and excessive shear in reactors causes callus generation and loss of productivity [165]. Several modified and novel designs have been proposed recently. A common feature of these reactors is that the roots are immobilised on supports within the vessel.

Jung and Tepfer [155] cultured *Calystegia sepium* hairy roots in a 30-litre stirred vessel containing a stainless-steel basket for attachment of the roots. Taya et al. [162] report culture of horseradish hairy-roots immobilised on polyurethane foam in a column reactor. The foam stood vertically in the reactor and air was pumped into the vessel at the bottom. As well as submerged culture, trickling-column operation and periodic filling and withdrawal of medium were tested. Growth of carrot hairy-roots and the relationship between growth rate, oxygen mass-transfer coefficient and reactor design have been investigated using three different bioreactors: air-driven, rotating-drum, and modified stirred-tank [167]. Growth in the air-driven vessel was slow; mass-transfer coefficients were significantly reduced in the presence of 10 $kg\,m^{-3}$ roots. Roots in the rotating drum were damaged until a polyurethane-foam sheet was attached to the inner wall of the rotating drum to provide protection and support. When 10 $kg\,m^{-3}$ roots were immobilised on the foam, mass-transfer coefficients increased significantly compared with those measured in the absence of cells. In the stirred-tank reactor, a stainless-steel mesh was installed to separate roots from the turbine impeller; mass transfer in this vessel was high and unaffected by root growth. In most reactors, oxygen supply to immobilised hairy-roots depends not only on bulk gas-liquid mass-transfer but also on diffusion into the tangled root mass.

A stirred reactor in which roots were isolated from the impeller by wire mesh was also tested by Hilton and Rhodes [168]. In this case a cylindrical mesh cage was used as support matrix for *Datura stramonium* hairy roots; this allowed a more even distribution of biomass up the length of the vessel. With continuous feeding of medium, packing densities reached 70% (drained weight per volume); however, mixing under these conditions was not ideal so that localised regions of low oxygen and nutrient levels are likely to have been formed.

One solution to the problem of oxygen transfer to hairy roots is use of mist reactors. In these vessels roots are not submerged in nutrient medium but are sprayed with fine droplets; the liquid drains through the biomass and is collected for recycle. Hairy roots respond well to direct exposure to air; Wilson et al. [169] have compared performance of *Datura stramonium* hairy roots in droplet- and submerged-culture reactors. A 7-d lag phase observed in submerged culture was eliminated in the mist reactor and final specific hyoscyamine levels were slightly higher.

In addition to biomass distribution and mass transfer there are other considerations affecting feasibility of commercial-scale hairy-root culture. These have been identified by Wilson et al. [169] as:

a) aseptic transfer of roots between inoculum and growth vessels; and
b) harvesting of the biomass.

Because of the tangled nature of root cultures, inoculation of large-scale production vessels is difficult to achieve aseptically. Roots grown on wire trellises or polyurethane foam are also difficult to remove from the support either for inoculation of a larger reactor or for product extraction. One solution to the problem of biomass harvesting is a collapsible root-immobilisation matrix described in a recent patent application by Wilson et al. [170]. A series of vertical stainless-steel rods make up the matrix; even after roots grow over the wires, the rods can be easily withdrawn by sliding them out of the root ball. This system has been tested with hairy roots in a 500-litre pilot-scale mist reactor.

4.2 Embryos

Mass culture of somatic embryos can be used for either plant propagation or production of phytochemicals.

Somatic embryogenesis is a proven means for producing large numbers of plants in vitro. Suspension cultures in which embryos float freely in the medium are especially amenable to large-scale mechanical handling. Somatic embryos of some plants do not respond well in liquid-culture; nevertheless, propagation using embryo suspensions has been demonstrated for vegetables such as celery and carrot and for several other species which readily form embryos in a liquid medium [171–173]. The ability to produce and grow embryos on a large scale is necessary for commercial application of this technology. Theoretically, large bioreactors would not be necessary since each embryo develops into an entire plant and more than 100000 embryos can be cultivated in a few litres of medium [174, 175]. In practice, however, embryo cultures contain a heterogeneous mixture of cells, cell clusters, embryos in various states of development and deformed embryos. Luckner and Diettrich [176] report that many *Digitalis lanata* embryos at the bipolar, heart-shaped or torpedo stage developed into plantlets with roots and leaves, but fewer than 1% developed into normal plants after further cultivation. Procedures for identifying viable embryos and separating them from the remainder of the culture are required.

Another use of embryos is direct sowing in the field; each embryo is a potential seedling and can be used as a seed substitute. Encapsulation of somatic embryos in gel at a suitable stage of development prevents premature desiccation and provides growth regulators which control seedling development. The benefits of artificial seeds include elimination of seed-borne diseases and development of improved strains in a relatively short time. Plant Genetics, Inc. in California has developed techniques for artificial seeds of celery and alfalfa [177]. Somatic embryos may also prove useful for long-term storage such as in germplasm banks.

Secondary metabolites are synthesised by somatic embryos; it is possible that high levels are produced because of the effects of cellular organisation. When torpedo-stage and older embryos produced in celery suspension cultures were analysed for production of flavour compounds, higher levels of secondary products accompanied increased differentiation [178]. The secondary-product composition of the embryo culture was comparable to that of the intact plant. Synthesis of cardiac glycosides by embryos of *Digitalis lanata* has been studied by Luckner and co-workers [173, 176, 179]. Large numbers of embryoids are formed at low auxin levels and can be grown as stage-I-globules in nutrient medium. After dilution and irradiation with light, the embryos develop into stage-II globules and start to accumulate cardenolides [173]. Cardenolide accumulation is highly sensitive to photoperiod and energy of irradiation [179].

Globules and developed embryos usually range in size from 0.5 to 1.5 mm and are well suited for culture in reactors. Airlift and stirred reactors were tested by Chen et al. [180] for mass-culture of alfalfa embryos; these studies were aimed at determining which operating parameters affect embryo formation. In the airlift vessel cell viability remained constant at about 60%; in the stirred reactors viability declined to less than 30% presumably due to shear effects. However embryogenesis was inhibited in the airlift vessel, especially at high oxygen levels. A 1-litre hanging-stirrer-bar reactor proved to be the most suitable configuration; viability was not adversely affected and embryo yields were enhanced. Depending on size, between 30–80% of the potential embryos produced in this reactor germinated normally on hormone-free medium and developed into plantlets. The resulting alfalfa plants were uniform and possessed normal morphology.

Development of stage-II globules and late-embryo structures of *Digitalis lanata* has been carried out in a 5-litre internal-loop airlift vessel irradiated by two high-pressure mercury lamps [181]. Cardenolide content of the embryos depended on the cell density and developmental stage at the time of inoculation, as well as on intensity of irradiation and nutrient composition. Concentrations of oxygen and carbon dioxide in the gas used to agitate the embryo suspension also affected growth and product synthesis.

Long periods of time are sometimes required for development of embryos from suspension culture; this limits their application for secondary-metabolite production. To alleviate this problem, immobilisation of embryonic cells has been investigated so that their biosynthetic capacity can be used in reactors for longer periods of time [182]. Cells of an embryonic strain of alfalfa were immobilised in polyurethane foam; embryoids were visible after 10d and filled the foam matrix. The ability of the immobilised embryos to carry out biotransformation reactions was also verified.

4.3 Shoots, Buds and Plantlets

Like embryo culture, large-scale plantlet culture can be used for either mass propagation or secondary-metabolite production. Micropropagation of many plant species by shoot-culture techniques on agar is well established; however organs can also be formed from suspended plant-cells. Requirements for shake-flask and bioreactor cultivation of differentiated plants are being investigated.

Conventional shoot propagation is labour intensive, expensive, and requires thousands of containers to produce a large number of plants. Use of submerged culture in reactors to propagate plants is potentially cost-effective, and roots and shoots can be initiated in the same vessel. Reactor-culture of ornamental plants has been demonstrated by Takayama and Misawa [183]. Regenerated buds of *Begonia* × *hiemalis* were grown in 3- and 10-litre air-sparged vessels; a maximum 22-fold increase in biomass was achieved. Growth in both these reactors was lower than in shake flasks; root growth was also severely inhibited. It was concluded that the major obstacle to use of reactors for large-scale plant propagation is injury to shoots caused by vigorous aeration. Damage was reduced when shoots adhered to the internals of the vessel and were thus protected from shear; this suggests that some form of immobilisation would be beneficial. Shoots removed from the medium were very turgid.

Further studies with a 2-litre rectangular-shaped air-driven reactor were carried out by Park et al. [184]. With young shoots of *Artemisia annua*, after 29-d growth the entire vessel volume was packed with shoots; this represented an 8-fold increase in biomass. The feasibility of direct root induction was also tested. After transferring shoots to the reactor containing low-glucose rooting medium, about 20% of the shoots produced roots in 32d; however the vessel became packed with shoots due to growth and circulation was impeded.

Alkaloid production in shoot cultures has been reviewed by Heble [185]. Medicinal compounds reported to accumulate at high levels in shoot cultures include those produced by *Atropa belladonna*, *Rauwolfia serpentina*, *Catharanthus roseus*, *Papaver bracteatum*, *Cinchona* sp., *Digitalis* sp. and *Dioscorea composita*. Reactor culture of alkaloid-producing shoots has been limited. Digitoxin synthesis in shoot-forming cultures of *Digitalis purpurea* is described by Hagimori et al. [186]; shoots were grown in a 3-litre air-sparged reactor with and without mechanical agitation. At the start of culture, plantlets circulated freely in the reactor; as their size increased due to generation of new leaves they floated and accumulated at the surface and after 15d the vessel was filled with plantlets. This high biomass density prevented uniform illumination of the culture with the result that cell digitoxin and chlorophyll contents depended on distance from the reactor wall. Mechanical agitation at 100 rpm had little effect on behaviour of the culture compared to when aeration only was used. Aeration rate had a strong influence on growth and digitoxin formation. Digitoxin content in the reactor-cultured shoots was a little lower than in shoots grown in shake flasks; light deficiency in the interior of the vessel was considered the most likely explanation.

5 Photoautotrophic Plant Cells

Photoautotrophic metabolism is not usually expressed in cultured plant cells because they are grown heterotrophically in dark or dim light conditions in the presence of an organic carbon source. However, photoautotrophic growth of plant cells in vitro has been achieved for about 23 different species [187]; these cultures provide new material for research on photosynthesis and chloroplast development and are also being considered as a way of increasing secondary metabolite production in tissue culture. Detailed descriptions of photoautotrophic cultures and their characteristics can be found in the reviews of Hüsemann [188], Hüsemann et al. [189] and Widholm [190]. Potential application of green hairy-roots of *Tagetes* and *Bidens* species for secondary metabolite production is discussed by Flores et al. [191].

In the laboratory, plant cells can be induced to develop photoautotrophic metabolism in two-tier culture flasks containing $K_2CO_3-KHCO_3$ buffer which produces a CO_2 atmosphere [192]. Formation of chlorophyll and fully functional chloroplasts in cultured plant cells has been shown [193] to involve:

a) lowering the sugar content and simultaneously increasing the CO_2 partial-pressure above ambient level;
b) preventing oxygen and ethylene accumulation;
c) providing high light intensities (6000–8000 lux); and
d) maintaining a balanced ratio of auxin/cytokinin in the medium.

When photoautotrophism is achieved, it is usually in sugar-free medium but in the presence of air containing 1–2% v/v CO_2. The reason why most cultured plant-cells are only capable of sustained photoautotrophic growth at highly elevated CO_2 partial pressures is still unknown; however, sustained photoautotrophic growth under ambient CO_2 concentrations has been demonstrated for selected cell lines of *Arachis hypogaea, Daucus carota* [194] and *Gossypium hirsutum* [190].

Since some enzymes involved in secondary synthesis are associated with chloroplasts and chloroplast membranes, it is expected that production rates of these compounds will be enhanced if chloroplasts are present. Comparative measurements of secondary-metabolite accumulation in heterotrophic, photo-mixotrophic and photoautotrophic cell have shown that green cell-cultures are able to express products partially or totally synthesised by chloroplast-localised enzymes. Systems in which greening enhances secondary production include *Lupinus polyphyllus* [195], *Morinda lucida* [196], *Solanum dulcamara* [197] and *Solanum laciniatum* [198]. In many cases, however, transition from heterotrophic to photoautotrophic conditions is not sufficient to induce secondary metabolism characteristics of leaves, and production of commercially-interesting secondary metabolites is either unchanged or decreased in photoautotrophic cells. For example, in *Nicotiana tabacum*, nicotine and other secondary metabolites occur mainly in heterotrophic rather than photoautotrophic cells, although N-methyl-nicotinic acid formation is totally restricted to photoautotrophic cultures [199, 200]. Hagimori et al. [201] found that digitoxin content in undifferentiated photoauto-trophic *Digitalis purpurea* cells was the same as in photomixotrophic cultures; in shoots, photoautotrophism reduced the alkaloid content to 10% of that in photomixotrophic cultures. Photoautotrophic suspensions of *Catharanthus roseus* synthesise very low levels of vindoline and dimeric alkaloids [202]; production of these compounds is drastically increased when heterotrophic conditions are restored [189].

For those compounds produced in quantity in photoautotrophic cells, large-scale production depends on design of suitable reactors. Important considerations for maintenance of photoautotrophic cultures include provision of elevated levels of CO_2, reduced levels of O_2, and adequate light intensity.

Continuous culture of photosynthetic plant-cell suspensions was demonstrated by Dalton [203]. *Spinacia oleracea* cells were cultured photoautotrophically at several steady states for more than 7 months in a 1.7-litre stirred reactor with dissolved-oxygen control and carbon-dioxide addition. Illumination was provided with 4–6 fluorescent lights; higher light intensities were obtained using a curved reflector of aluminium foil. The results from these experiments suggested that light intensity limited biomass accumulation and biosynthesis. Batch and fed-batch reactor cultures of green *Asparagus officinalis* and *Ocimum basilicum* cells were subsequently carried out [204].

Photoautotrophic growth of *Nicotiana tabacum* was achieved in a 5-litre stirred reactor with an incident light intensity of 8000 lux and an aeration gas containing 1% CO_2, 14% O_2, and the remainder N_2 [205]. Enrichment with 1% CO_2 without a simultaneous reduction in O_2 partial pressure did not induce photoautotrophism. Peel [206] cultured photoautotrophic cells of *Asparagus officinalis* at steady state in a small turbidostat reactor for 21d. Hüsemann [207] reports use of a 1.5-litre inverted-flask bubble-column reactor for photoautotrophic culture of *Chenopodium rubrum* in the presence of 2% CO_2 and 8000 lux external light intensity. Chlorophyll and protein contents and activities of ribulose bisphosphate carboxylase and phosphoenolpyruvate carboxylase enzymes were reduced compared with cultures grown in shake flasks. With the same equipment, Hüsemann [208] used continuous-culture photoautotrophic conditions to show that multiplication and differentiation of chloroplasts keep pace with cell division. Scale-up of suspended photoautotrophic *Chenopodium rubrum* cells in a 17-litre airlift vessel has been described by Fischer and Alfermann [209].

Uniform and adequate illumination of cells is an important problem in large-scale photoautotrophic culture and becomes more difficult in large reactors if an external light source is used. To overcome these problems, Ohta and Takata [210] have designed a 10-litre vessel in which light is transmitted by optical fibres. The light source is a xenon lamp or solar-ray collector; quartz optical fibres coupled with polyacrylate resin fibres are inserted into the reactor through protective glass tubing. This system has been tested with photoheterotrophic cultures of liverwort. Provided that sufficient light intensity can be transmitted using optical fibres, devices providing an internal source of light are also likely to be successful with photosynthesising plant-cell cultures.

6 Conclusions and Future Outlook

The newest developments in plant-cell reactor design are those aimed at culture of differentiated organs and tissue; this leaves many unsolved problems with reactor culture of high-density suspended plant-cells. Although air-driven bioreactors work well at suspended-cell concentrations less than 12–15 kg m^{-3} dry weight, mixing becomes limiting at higher biomass densities. Stirred reactors have more potential for large-scale suspended-cell processes because they are easily able to provide adequate mixing and mass transfer; however, the shear sensitivity of many plant cells restricts application of standard configurations. Research into inducing or selecting for shear tolerance is a worthwhile approach so that the range of application of agitated systems can be broadened. Alternatively, design of impellers which maintain high levels of mixing and mass transfer but reduce the intensity of shear may provide a satisfactory solution.

Design of reactors for immobilised plant-cells and organised tissue is at an early stage. Many standard reactor configurations appear adequate for immobilised cell culture. For hairy roots, immobilisation of the biomass to ensure uniform distribution in the reactor is important; efficient ways of harvesting the cells must then be found. The general applicability of embryos, shoots and plantlets for secondary-metabolite production is yet to be demonstrated. However, protection from shear and provision of adequate light are important considerations in reactor design.

Engineering analysis of large-scale plant-cell culture must go hand-in-hand with more fundamental studies of plant-cell metabolism. Requirements for growth of cells may be very different from those for maximum productivity so that criteria for reactor design will vary. As an example, aiming for $k_L a_L$ values which do not limit growth seems appropriate; however secondary production could be enhanced under oxygen-limited conditions, or high air flow rates may deplete the culture of vital components such as carbon dioxide and ethylene. Uncertainty about culture conditions would be reduced if current studies of pathway regulation in plant cells can provide a formula for high product yields. In many ways, further refinement of reactor design for plant-cell culture depends on the outcome of research into how secondary synthesis can be triggered.

7 Nomenclature

A_d	downcomer cross-sectional area
a_D	interfacial area per unit dispersed volume
a_L	interfacial area per unit volume unaerated liquid
A_r	riser cross-sectional area
C	oxygen concentration within particle
C^*	oxygen concentration in liquid in equilibrium with air
C_e	oxygen concentration at surface of particle
C_L	oxygen concentration in liquid
C_s	solids concentration
$\mathcal{D}$	diffusivity
D_b	bubble diameter
D_c	column diameter
D_d	downcomer diameter
$\mathcal{D}_e$	effective diffusion coefficient
D_i	impeller diameter
$\mathcal{D}_L$	diffusivity in liquid
D_r	riser diameter
D_T	tank diameter
f_m	dimensionless mixing group

g gravitational acceleration
H_D gas-liquid dispersion height
H_d height of the downcomer
H_L liquid height
H_r height of the riser
K power-law consistency index
k empirical constant in Eq. (9)
K_B frictional loss coefficient for bottom of airlift reactor
k_L liquid-film mass transfer coefficient
K_m Michaelis-Menten constant
K_T frictional loss coefficient for top of airlift reactor
k_w resistance coefficient in Eq. (32)
n power-law flow behaviour index
N_i impeller rotational speed
OTR oxygen transfer rate
P power input
ΔP_{Fd} frictional pressure drop in the downcomer
ΔP_{Fr} frictional pressure drop in the riser
q_{O_2} observed specific oxygen consumption rate
R radius of particle
r radial distance measured from centre of particle
Re_i impeller Reynolds number
r_{O_2} volumetric oxygen consumption rate
t_c circulation time
t_{mt} mass transfer time constant
t_{mx} mixing time constant
t_{rxn} oxygen consumption time constant
u_B bubble rise velocity
u_G superficial gas velocity
u_{Gr} superficial gas velocity in the riser
u_L liquid velocity
u_{Ld} superficial liquid velocity in the downcomer
u_{Lr} superficial liquid velocity in the riser
V_L liquid volume
v_{max} maximum reaction velocity
X cell concentration
β exponent in Eq. (26)
$\dot{\gamma}$ shear rate
$\dot{\gamma}_{av}$ average shear rate
$\dot{\gamma}_{max}$ maximum shear rate
ε local energy dissipation rate
ε_d gas hold-up in the downcomer
ε_r gas hold-up in the riser
ε_T total gas hold-up
η eddy size
λ characteristic material time

μ_{app} apparent viscosity
μ_g gas viscosity
μ_L liquid viscosity
ν fluid kinematic viscosity
ϱ_L liquid density
σ surface tension
τ shear stress
τ_0 yield stress
τ_w wall shear stress

8 References

1. Fowler MW (1983) In: Mantell SH, Smith H (eds) Plant biotechnology. Cambridge University Press, Cambridge, p 3
2. Misawa M (1985) Adv Biochem Eng/Biotechnol 31: 59
3. Kurz WGW, Constabel F (1985) CRC Crit Rev Biotechnol 2: 105
4. Rhodes MJC, Robins RJ, Hamill J and Parr AJ (1986) NZ J Technol 2: 59
5. Kato A, Kawazoe S and Soh Y (1978) J Ferment Technol 56: 224
6. Roels JA, van den Berg J and Voncken RM (1974) Biotechnol Bioeng 16: 181
7. Charles M (1978) Adv Biochem Eng 8: 1
8. Metz B, Kossen NWF and van Suijdam JC (1979) Adv Biochem Eng 11: 103
9. Wagner F, Vogelmann H (1977) In: Barz W, Reinhard E, Zenk MH (eds) Plant tissue culture and its bio-technological applications. Springer, Berlin Heidelberg New York, p 245
10. Tanaka H (1982) Biotechnol Bioeng 24: 425
11. Scragg AH, Allan EJ, Bond PA and Smart NJ (1986) In: Morris P, Scragg AH, Stafford A and Fowler MW (eds) Secondary metabolism in plant cell cultures. Cambridge University Press, Cambridge, p 178
12. Vogelmann H (1981) In: Moo-Young M, Robinson CW and Vezina C (eds) Advances in biotechnology, vol 1. Pergamon, New York, p 117
13. Laine J and Kuoppamäki R (1979) Ind Eng Chem Process Des Dev 18: 501
14. Takayama S, Misawa M, Ko K and Misato T (1977) Physiol Plant 41: 313
15. Sahai OP and Shuler ML (1982) Can J Bot 60: 692
16. Street HE (1973) In: Milborrow BV (ed) Biosynthesis and its control in plants. Academic, London, p 95
17. Tanaka H, Semba H, Jitsufuchi T and Harada H (1988) Biotechnol Lett 10: 485
18. Metzner AB and Otto RE (1957) AIChE J 3: 3
19. Hooker BS, Lee JM and An G (1990) Biotechnol Bioeng 35: 296
20. Scragg AH, Allan EJ and Leckie F (1988) Enzyme Microb Technol 10: 361
21. Hooker BS, Lee JM and An G (1989) Enzyme Microb Technol 11: 484
22. Allan EJ, Scragg AH and Pugh K (1988) J Plant Physiol 132: 176
23. Meijer J (1990) PhD Thesis, Delft University of Technology, The Netherlands
24. Midler M and Finn RK (1966) Biotechnol Bioeng 8: 71
25. Chen H, Wang J-J and Liu Y-G (1990) In: Abstracts VIIth Internat Congr Plant Tiss Cell Cult, Amsterdam, 24–29 June 1990, p 341
26. Cherry RS and Papoutsakis ET (1986) Bioprocess Eng 1: 29
27. Tramper J, Williams JB, Joustra D and Vlak JM (1986) Enzyme Microb Technol 8: 33
28. Croughan MS, Sayre ES and Wang DIC (1989) Biotechnol Bioeng 33: 862
29. Handa-Corrigan A, Emery AN and Spier RE (1989) Enzyme Microb Technol 11: 230
30. Kunas KT and Papoutsakis ET (1990) Biotechnol Bioeng 36: 476
31. Murhammer DW and Goochee CF (1990) Biotechnol Prog 6: 391

32. Jöbses I, Martens D and Tramper J (1991) Biotechnol Bioeng 37: 484
33. Bavarian F, Fan LS and Chalmers JJ (1991) Biotechnol Prog 7: 140
34. Chalmers JJ and Bavarian F (1991) Biotechnol Prog 7: 151
35. Cherry RS and Kwon K-Y (1990) Biotechnol Bioeng 36: 563
36. Tanaka H (1981) Biotechnol Bioeng 23: 1203
37. Ulbrich B (1986) In: Korhola M, Tuompo H and Kauppinen V (eds) Foundation for biotechnical and industrial fermentation research 4. Proc 7th Conf on Global Impacts of Appl Microbiol, p 147
38. Hong YC, Labuza TP and Harlander SK (1989) Biotechnol Prog 5: 137
39. Smart NJ and Fowler MW (1984) Appl Biochem Biotechnol 9: 209
40. Hegarty PK, Smart NJ, Scragg AH and Fowler MW (1986) J Exp Bot 37: 1911
41. Scragg AH, Cresswell RC, Ashton S, York A, Bond PA and Fowler MW (1989) Enzyme Microb Technol 11: 329
42. Scragg AH, Ashton S, York A, Bond P, Stepan-Sarkissian G and Grey D (1990) Enzyme Microb Technol 12: 292
43. Goldstein WE, Lasure LL and Ingle MB (1980) In: Staba EJ (ed) Plant tissue culture as a source of biochemicals. CRC Press, Boca Raton, USA, p 191
44. Tanaka H (1987) Process Biochem 22: 106
45. Noguchi M, Matsumoto T, Hirata Y, Yamamoto K, Katsuyama A, Kato A, Azechi S and Kato K (1977) In: Barz W, Reinhard E and Zenk MH (eds) Plant tissue culture and its bio-technological applications. Springer, Berlin Heidelberg New York, p 85
46. Curtin ME (1983) Bio/Technol 1: 649
47. Fujita Y (1988) In: Bock G and Marsh J (eds) Applications of plant cell and tissue culture. Ciba foundation symp 137, Wiley, Chichester, p 228
48. Westphal K (1990) In: Progress in plant cellular and molecular biology. Proc VIIth Internat Congr Plant Tiss Cell Cult, Amsterdam, 24–29 June 1990, Kluwer Academic, Dordrecht, p 601
49. Leckie F, Scragg AH and Cliffe KC (1990) In: Progress in plant cellular and molecular biology. Proc VIIth Internat Congr Plant Tiss Cell Cult, Amsterdam, 24–29 June 1990, Kluwer Academic, Dordrecht, p 689
50. Chisti MY (1989) Airlift bioreactors, Elsevier, London
51. Nagata S (1975) Mixing: principles and applications, John Wiley, New York
52. Bello RA, Robinson CW and Moo-Young M (1984) Can J Chem Engn 62: 573
53. Holland FA and Chapman FS (1966) Liquid mixing and processing in stirred tanks. Reinhold, New York
54. Metzner AB and Taylor JS (1960) AIChE J 6: 109
55. Taguchi H (1971) Adv Biochem Eng 1: 1
56. Johnson DN and Hubbard DW (1974) Biotechnol Bioeng 16: 1283
57. Norwood KW and Metzner AB (1960) AIChE J 6: 432
58. Wang DIC and Fewkes RCJ (1977) In: Development in industrial microbiology, vol 18. Proc 33rd Gen Meet Soc Ind Microbiol, Jekyll Island, Georgia, USA, Aug 14–20 1976, p 39
59. Blakebrough N and Sambamurthy K (1966) Biotechnol Bioeng 8: 25
60. Calderbank PH and Moo-Young MB (1959) Trans IChE 37: 26
61. Rousseau I and Bu'Lock JD (1980) Biotechnol Lett 2: 475
62. Weiland P (1984) Ger Chem Eng 7: 374
63. Chisti MY and Moo-Young M (1987) Chem Eng Comm 60: 195
64. Chisti MY, Halard B and Moo-Young M (1988) Chem Eng Sci 43: 451
65. Chakravarty M, Singh HD, Baruah JN and Iyengar MS (1974) Indian Chem Engr 16: 17
66. Popovic M and Robinson CW (1988) Biotechnol Bioeng 32: 301
67. Nishikawa M, Kato H and Hashimoto K (1977) Ind Eng Chem Process Des Dev 16: 133
68. Allen DG and Robinson CW (1989) Biotechnol Bioeng 34: 731
69. Kessell RHJ and Carr AH (1972) J Exp Bot 23: 996
70. Payne GF, Shuler ML and Brodelius P (1987) In: Lydersen BK (ed) Large scale cell culture technology. Hanser, Munich, p 193

71. Kato A, Shimizu Y and Nagai S (1975) J Ferment Technol 53: 744
72. Pareilleux A and Vinas R (1983) J Ferment Technol 61: 429
73. Mavituna F (1988) In: Pais MSS, Mavituna F, Novais JM (eds) Plant cell biotechnology. NATO ASI Series H18, Springer, Berlin, Heidelberg, New York, p 1
74. Kobayashi Y, Fukui H and Tabata M (1989) Plant Cell Rep. 8: 255
75. van't Riet K (1979) Ind Eng Chem Process Des Dev 18: 357
76. Schügerl K (1981) Adv Biochem Eng 19: 71
77. Deindoerfer FH and Gaden EL (1955) Appl Microbiol 3: 253
78. Yagi H and Yoshida F (1975) Ind Eng Chem Process Des Dev 14: 488
79. Perez JF and Sandall OC (1974) AIChE J 20: 770
80. Chisti MY and Moo-Young M (1988) Chem Eng J 38: 149
81. Bello RA, Robinson CW and Moo-Young M (1985) Biotechnol Bioeng 28: 369
82. Kawase Y and Moo-Young M (1986) Chem Eng Commun 40: 67
83. Chisti MY, Fujimoto K and Moo-Young M (1987) In: Ho CS and Oldshue JY (eds) Biotechnology processes – scale-up and mixing. AIChE, New York, p 72
84. Popovic MK and Robinson CW (1989) AIChE J 35: 393
85. Oldshue JY (1966) Biotechnol Bioeng 8: 3
86. Wichterle K, Kadlec M, Zak L and Mitschka P (1984) Chem Eng Commun 26: 25
87. Henzler H-J (1980) Chem Ing Tech 52: 643
88. Tramper J, Joustra D and Vlak JM (1987) In: Webb C and Mavituna F (eds) Plant and animal cells: process possibilities. Ellis Horwood, Chichester, p 125
89. van't Riet K and Tramper H (1991) Basic bioreactor design. Marcel Dekker, New York
90. Kossen NWF (1984) Proc 3rd Eur Fed Biotechnol Congr, 10–14 September 1984, Munich, vol 4, Weinheim VCH, p 257
91. Chisti MY and Moo-Young M (1988) Biotechnol Bioeng 31: 487
92. Townsley PM, Webster F, Kutney JP, Salisbury P, Hewitt G, Kawamura N, Choi L, Kurihara T and Jacoli GG (1983) Biotechnol Lett 5: 13
93. Drapeau D, Blanch HW and Wilke CR (1986) Biotechnol Bioeng 28: 1555
94. Scragg AH and Fowler MW (1985) In: Vasil IK (ed) Cell culture and somatic cell genetics of plants, vol 2. Academic, Orlando, p 103
95. Wang C-J and Staba EJ (1963) J Pharm Sci 52: 1058
96. Bond PA, Hegarty P and Scragg AH (1987) Proc 4th Eur Congr Biotechnol 1987, vol 2. Neijssel OM, van der Meer RR and Luyben KChAM (eds), Elsevier, Amsterdam, p 440
97. Kawase Y and Moo-Young M (1990) Bioprocess Eng 5: 169
98. Bourne JR and Butler H (1965) AIChE-IChE Symp Ser 10: 89
99. Ulbrich B, Wiesner W and Arens H (1985) In: Neumann K-H, Barz W and Reinhard E (eds) Primary and secondary metabolism in plant cell cultures. Springer, Berlin, Heidelberg, New York, p 293
100. Tanaka H, Nishijima F, Suwa M and Iwamoto T (1983) Biotechnol Bioeng 25: 2359
101. Brodelius P, Deus B, Mosbach K and Zenk MH (1979) FEBS Lett 103: 93
102. Alfermann AW, Schuller I and Reinhard E (1980) Planta Med 40: 218
103. Veliky IA and Jones A (1981) Biotechnol Lett 3: 551
104. Furuya T, Yoshikawa T and Taira M (1984) Phytochem 23: 999
105. Jirku V, Macek T, Vanek T, Krumphanzl V and Kubanek V (1981) Biotechnol Lett 3: 447
106. Lindsey K, Yeoman MM, Black GM and Mavituna F (1983) FEBS Lett 155: 143
107. Shuler ML, Sahai OP and Hallsby GA (1983) Ann NY Acad Sci 413: 373
108. Lindsey K and Yeoman MM (1983) In: Mantell SH and Smith H (eds) Plant biotechnology. Cambridge University Press, Cambridge, p 39
109. Robins RJ, Parr AJ, Richards SR and Rhodes MJC (1986) In: Morris P, Scragg AH, Stafford A and Fowler MW (eds) Secondary metabolism in plant cell cultures. Cambridge University Press, Cambridge, p 162
110. Brodelius P (1984) In: Vasil IK (ed) Cell culture and somatic cell genetics of plants, vol 1. Academic, New York, p 535

111. Brodelius P (1986) In: Evans DA, Sharp WR and Ammirato PV (eds) Handbook of plant cell culture, vol 4. Macmillan, New York, p 287
112. Novais JM (1988) In: Pais MSS, Mavituna F and Novais JM (eds) Plant cell biotechnology. NATO ASI Ser H18, Springer, Berlin, Heidelberg, New York, p 353
113. Rosevear A (1988) In: King RD and Cheetham PSJ (eds) Food biotechnology, vol 2. Elsevier, London, p 83
114. Hulst AC and Tramper J (1989) Enzyme Microb Technol 11: 546
115. Brodelius P and Nilsson K (1980) FEBS Lett 122: 312
116. Ketel DH, Hulst AC, Gruppen H, Breteler H and Tramper J (1987) Enzyme Microb Technol 9: 303
117. Hulst AC, Tramper J, Brodelius P, Eijkenboom LJC and Luyben KChAM (1985) J Chem Tech Biotechnol 35B: 198
118. Ayabe S-I, Iida K and Furuya T (1986) Phytochem 25: 2803
119. Prenosil JE, Hegglin M, Baumann TW, Frischknecht PM, Kappeler AW, Brodelius P and Haldimann D (1987) Enzyme Microb Technol 9: 450
120. Holden MA, Hall RD, Lindsey K and Yeoman MM (1987) In: Webb C and Mavituna F (eds) Plant and animal cells: process possibilities. Ellis Horwood, Chichester, p 45
121. Ishida BK (1988) Plant Cell Rep 7: 270
122. Barnabas NJ and David SB (1988) Biotechnol Lett 10: 593
123. Brodelius P and Nilsson K (1983) Eur J Appl Microbiol Biotechnol 17: 275
124. Robins RJ and Rhodes MJC (1986) Appl Microbiol Biotechnol 24: 35
125. Mavituna F, Wilkinson AK, Williams PD and Park JM (1987) In: Moody GW and Baker PB (eds) Bioreactors and biotransformations. Elsevier, London, p 26
126. Kim DJ and Chang HN (1990) Biotechnol Bioeng 36: 460
127. Ford JR, Lambert AH, Cohen W and Chambers RP (1972) Biotechnol Bioeng Symp 3: 267
128. Hulst AC, Meyer MMT, Breteler H and Tramper J (1989) Appl Microbiol Biotechnol 30: 18
129. Mavituna F, Park JM, Wilkinson AK and Williams PD (1987) In: Webb C and Mavituna F (eds) Plant and animal cells: process possibilities. Ellis Horwood, Chichester, p 92
130. Furusaki S, Nozawa T, Isohara T and Furuya T (1988) Appl Microbiol Biotechnol 29: 437
131. Pras N, Hesselink PGM, Guikema WM and Malingré TM (1989) Biotechnol Bioeng 33: 1461
132. Kargi F, Ganapathi B and Maricic K (1990) Biotechnol Prog 6: 243
133. Fuller KW (1984) Chem Ind 3 Dec, p 825
134. Dainty AL, Goulding KH, Robinson PK, Simpkins I and Trevan MD (1985) Trends in Biotechnol 3: 59
135. Hoekstra SS, Harkes PAA, Verpoorte R and Libbenga KR (1990) Plant Cell Rep 8: 571
136. Tsoulpha P and Doran PM (1991) J Biotechnol 19: 99
137. Fuller KW and Bartlett DJ (1982) PCT Patent Application PCT/GB81/00186
138. Panda AK, Mishra S, Bisaria VS and Bhojwani SS (1989) Enzyme Microb Technol 11: 386
139. Kobayashi Y, Fukui H and Tabata M (1987) Plant Cell Rep 6: 185
140. Kobayashi Y, Fukui H and Tabata M (1988) Plant Cell Rep 7: 249
141. Archambault J, Volesky B and Kurz WGW (1989) Biotechnol Bioeng 33: 293
142. Archambault J, Volesky B and Kurz WGW (1990) Biotechnol Bioeng 35: 702
143. Kurz WGW, Paiva NL and Tyler RT (1990) In: Progress in plant cellular and molecular biology. Proc VIIth Internat Congr Plant Tiss Cell Cult, Amsterdam, 24–29 June 1990, Kluwer Academic, Dordrecht, p 682
144. Tramper J, Wolters I and Verlaan P (1987) In: Laane C, Tramper J and Lilly MD (eds) Biocatalysis in organic media. Elsevier, Amsterdam, p 311
145. Buitelaar RM, Susaeta I and Tramper J (1990) In: Progress in plant cellular and molecular biology. Proc VIIth Internat. Congr Plant Tiss Cell Cult, Amsterdam, 24–29 June 1990, Kluwer Academic, Dordrecht, p 694

146. Haldimann D and Brodelius P (1987) Phytochem 26: 1431
147. Brodelius P (1984) Ann NY Acad Sci 434: 382
148. Vanek T and Macek T (1989) Planta Med 55: 680
149. Doran PM and Bailey JE (1986) Biotechnol Bioeng 28: 73
150. Bhandary SBR, Collin HA, Thomas E and Street HE (1969) Ann Bot 33: 647
151. Tabata M, Yamamoto H, Hiraoka N and Konoshima M (1972) Phytochem 11: 949
152. Flores HE (1987) In: LeBaron HM, Mumma RO, Honeycutt RC and Duesing JH (eds) Biotechnology in agricultural chemistry, ACS Symp Series 334, American Chemical Society, Washington DC, p 66
153. Aird ELH, Hamill JD and Rhodes MJC (1988) Plant Cell Tiss Organ Cult 15: 47
154. Parr AJ and Hamill JD (1987) Phytochem 26: 3241
155. Jung G and Tepfer D (1987) Plant Sci 50: 145
156. Christen P, Roberts MF, Phillipson JD and Evans WC (1989) Plant Cell Rep 8: 75
157. White PR (1934) Plant Physiol 9: 585
158. Rhodes MJC, Robins RJ, Hamill JD, Parr AJ and Walton NJ (1987) Internat Assoc Plant Tiss Cult Newsletter 53, Nov 1, p 2
159. Hamill JD, Parr AJ, Rhodes MJC, Robins RJ and Walton NJ (1987) Bio/Technol 5: 800
160. Doran PM (1989) Aust J Biotechnol 3: 270
161. Signs MW and Flores HE (1990) BioEssay 12: 7
162. Taya M, Yoyama A, Kondo O, Kobayashi T and Matsui C (1989) J Chem Eng Japan 22: 84
163. Sharp JM and Doran PM (1990) J Biotechnol 16: 171
164. Rhodes MJC, Hilton M, Parr AJ, Hamill JD and Robins RJ (1986) Biotechnol Lett 8: 415
165. Wilson PDG, Hilton MG, Robins RJ and Rhodes MJC (1987) In: Moody GW and Baker PB (eds) Bioreactors and biotransformations. Elsevier, London, p 38
166. Davioud E, Kan C, Hamon J, Tempé J and Husson H-P (1989) Phytochem 28: 2675
167. Kondo O, Honda H, Taya M and Kobayashi T (1989) Appl Microbiol Biotechnol 32: 291
168. Hilton MG and Rhodes MJC (1990) Appl Microbiol Biotechnol 33: 132
169. Wilson PDG, Hilton MG, Meehan PTH, Waspe CR and Rhodes MJC (1990) In: Progress in plant cellular and molecular biology. Proc VIIth Internat Congr Plant Tiss Cell Cult, Amsterdam, 24–29 June 1990, Kluwer Academic, Dordrecht, p 700
170. Wilson PDG, Hilton MG, Steer DC, Waspe CR, Robins RJ and Rhodes MJC (1989) PCT Patent Application PCT/GB89/00510
171. Konar RN, Thomas E and Street HE (1972) Ann Bot 36: 249
172. Al-Abta S and Collin HA (1978) New Phytol 80: 517
173. Kuberski C, Scheibner H, Steup C, Diettrich B and Luckner M (1984) Phytochem 23: 1407
174. Styer DJ (1985) In: Henke RR, Hughes KW, Constantin MJ and Hollaender A (eds) Tissue culture in forestry and agriculture. Plenum, New York, p 117
175. Cazzulino DL, Pedersen H, Chin C-K and Styer D (1990) Biotechnol Bioeng 35: 781
176. Luckner M and Diettrich B (1988) In: Constabel F and Vasil IK (eds) Cell culture and somatic cell genetics of plants, vol 5. Academic, San Diego, p 193
177. Redenbaugh K, Paasch BD, Nichol JW, Kossler ME, Viss PR and Walker KA (1986) Bio/Technol 4: 797
178. Al-Abta S, Galpin IJ and Collin HA (1979) Plant Sci Lett 16: 129
179. Scheibner H, Björk L, Schulz U, Diettrich B and Luckner M (1987) J Plant Physiol 130: 211
180. Chen THH, Thompson BG and Gerson DF (1987) J Ferment Technol 65: 353
181. Greidziak N, Diettrich B and Luckner M (1990) Planta Med 56: 175
182. Macek T, Vanek T and Binarova P (1989) Planta Med 55: 595
183. Takayama S and Misawa M (1981) Plant Cell Physiol 22: 461
184. Park JM, Hu W-S and Staba EJ (1989) Biotechnol Bioeng 34: 1209
185. Heble MR (1985) In: Neumann K-H, Barz W and Reinhard E (eds) Primary and secondary metabolism of plant cell cultures. Springer, Berlin, Heidelberg, New York, p 281

186. Hagimori M, Matsumoto T and Mikami Y (1984) Agric Biol Chem 48: 965
187. Hüsemann W, Amino S, Fischer K, Herzbeck H and Callis R (1990) In: Progress in plant cellular and molecular biology. Proc VIIth Internat Congr Plant Tiss Cell Cult, Amsterdam, 24–29 June 1990, Kluwer Academic, Dordrecht, p 373
188. Hüsemann W (1985) In: Vasil IK (ed) Cell culture and somatic cell genetics of plants, vol 2. Academic, Orlando, p 213
189. Hüsemann W, Fischer K, Mittelbach I, Hübner S, Richter G and Barz W (1989) In: Kurz WGW (ed) Primary and secondary metabolism of plant cell cultures II. Springer, Berlin, Heidelberg, New York, p 35
190. Widholm JM (1989) In: Kurz WGW (ed) Primary and secondary metabolism of plant cell cultures II. Springer, Berlin, Heidelberg, New York, p 3
191. Flores HE, Hoy MW and Pickard JJ (1987) Trends in biotechnol. 5: 64
192. Hüsemann W and Barz W (1977) Physiol Plant 40: 77
193. Hüsemann W (1988) In: Pais MSS, Mavituna F and Novais JM (eds) Plant cell biotechnology. NATO ASI Ser H18, Springer, Berlin, Heidelberg, New York, p 179
194. Bender L, Kumar A and Neumann K-H (1985) In: Neumann K-H, Barz W and Reinhard E (eds) Primary and secondary metabolism of plant cell cultures. Springer, Berlin, Heidelberg, New York, p 24
195. Wink M and Hartmann T (1980) Planta Med 40: 149
196. Igbavboa U, Sieweke H-J, Leistner E, Röwer I, Hüsemann W and Barz W (1985) Planta 166: 537
197. Emke A and Eilert U (1986) Plant Cell Rep 5: 31
198. Conner AJ (1987) Phytochem 26: 2749
199. Ikemeyer D and Barz W (1989) Plant Cell Rep 8: 479
200. Barz W, Ikemeyer D and Beimen A (1989) Planta Med 55: 594
201. Hagimori M, Matsumoto T and Mikami Y (1984) Plant Cell Physiol 25: 1099
202. Tyler RT, Kurz WGW and Panchuk BD (1986) Plant Cell Rep 3: 195
203. Dalton CC (1980) J Exp Bot 31: 791
204. Dalton CC and Peel E (1983) Prog Ind Microbiol 17: 109
205. Yamada Y, Imaizumi K, Sato F and Yasuda T (1981) Plant Cell Physiol 22: 917
206. Peel E (1982) Plant Sci Lett 24: 147
207. Hüsemann W (1982) Protoplasma 113: 214
208. Hüsemann W (1983) Plant Cell Rep 2: 59
209. Fischer U and Alfermann AW (1989) Planta Med 55: 686
210. Ohta Y and Takata T (1990) In: Abstracts VIIth Internat Congr Plant Tiss Cell Cult, Amsterdam, 24–29 June 1990, p 341
211. Schumpe A and Deckwer W-D (1987) Bioprocess Eng 2: 79
212. El-Temtamy SA, Khalil SA, Nour-El-Din AA and Gaber A (1984) Appl Microbiol Biotechnol 19: 376
213. Henzler H-J and Kauling J (1985) In: Proc 5th Eur Conf on Mixing, Würzburg, Germany, 10–12 June 1985, p 303

Expert Systems in Bioprocess Control: Requisite Features

Konstantin B. Konstantinov[1]*, Robert Aarts[2] and Toshiomi Yoshida[1]

[1] International Center of Biotechnology, Faculty of Engineering, Osaka University, Yamada oka, Suita shi, Osaka 565, Japan
[2] VTT, Biotechnical Laboratory, P.O. Box 202, SF-02151 Espoo, Finland

The development of intelligent control in biotechnical processes, which only a decade ago was considered an exciting, but obscure vision, has today become an area of intensive and realistic research. One of the keys to success in this field is the selection of an adequate software tool for building the intelligent system. The ideal tool must possess a large set of features, which reflect both the real-time nature of the control problem, and the peculiarities of the biotechnical system itself. Here, some of these features are introduced. In addition, the main concepts and trends in the field of expert control in biotechnical processes are discussed.

* To whom all correspondence should be addressed. Present address: Department of Chemical Engineering, University of Delaware, DE 19716, USA.

Advances in Biochemical Engineering
Biotechnology, Vol. 48
Managing Editor: A. Fiechter

1 Introduction

Because of the focus on formal analysis and synthesis, many other aspects of control system design have largely been disregarded in various application areas [1]. This has been the case with bioprocesses control, even though the limitations of the traditional control approaches in the field of biological systems have long been perceived. Nowadays, it is becoming clear that conventional control theory alone cannot provide the required platform for building high-performance systems for the control of bioprocesses. Consequently, a trend towards exploitation of new methods capable of manipulating and utilizing informal knowledge of the biological plant has emerged. Recent achievements in expert system (ES) technology have already stimulated research aimed at its application in the field of bioprocesses control [2].

However, when turning to the ES approach, the bioprocess engineer is likely to encounter a number of unexpected difficulties, which arise from the still limited experience in the field of expert control, the fundamental differences from traditional control methodology, and the development of the software system itself. One of the tasks which must be solved at an early stage of a project is the selection of an appropriate expert system development tool (ESDT), also called an "expert shell". This is crucial to the overall design process; an unsuitable tool may predestine the research to fruitless effort for years, while selection of the right one provides a reliable basis for rapid success. Unfortunately, the significance of this problem is still not sufficiently realized. There is a tendency for people to think of ESs as one thing, and to consider that any ESDT will suit their particular case [3]. In fact, current ES technology is leading to the creation of hundreds of products, which vary tremendously in size, purpose, capability and price. Since a commercial ESDT specialized for control of bioprocesses has not yet been produced, developers should make their choice only after careful screening of the tools currently available on the market. The product selected must provide a large set of features, many of which are non-trivial and are not supported by the more common ESDTs. These reflect both the real-time nature of the control problem, and the peculiarities of the biotechnical system itself.

The principal purpose of this paper is to introduce a set of ESDT features which are required in building intelligent systems for the control of bioprocesses. In addition, the basic structure, functions, and current limitations of ES technology in the context of problems encountered in the control of bioprocesses, are discussed.

2 Basic Structures, Functions and Implementation Schemes

2.1 Structures of Systems for Expert Control

There are many ways to incorporate expert knowledge into a process control system. Nevertheless, almost all possible cases can be reduced to two basic schemes for expert control, referred to as *direct* and *indirect* [1]. Direct systems are those

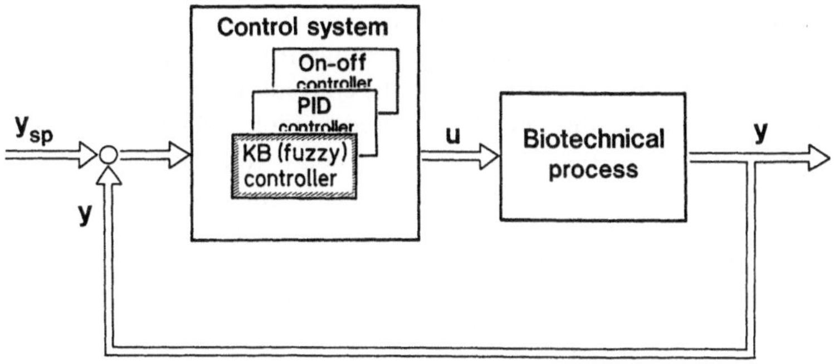

Fig. 1. Structure of a system for direct expert control. The knowledge-based (KB) controller works at the level of the standard controllers, such as PID or "on-off"

whose knowledge-based modules are involved in the control loop (Fig. 1). These modules operate at the level of standard PID controllers, and are useful in realizing more complicated, nonlinear control algorithms. Such modules are known mostly as fuzzy controllers, because they are often based on fuzzy logic. Recent achievements in neural networks provide another alternative for the design of such types of controllers. Generally, the direct expert control scheme, though useful in some cases, is limited to local and low-level problems. This means that although its applicability to the control of bioprocesses is possible, it is of limited significance.

The structure of a system for indirect expert control (also called "supervisory") is shown in Fig. 2 [4–6]. This is composed of two hierarchical levels providing clear distribution of the system functions. The standard set of control tasks, such as measurement, filtration, data acquisition, control, etc., are entrusted to the lower level, which represents just a conventional control system. Such systems

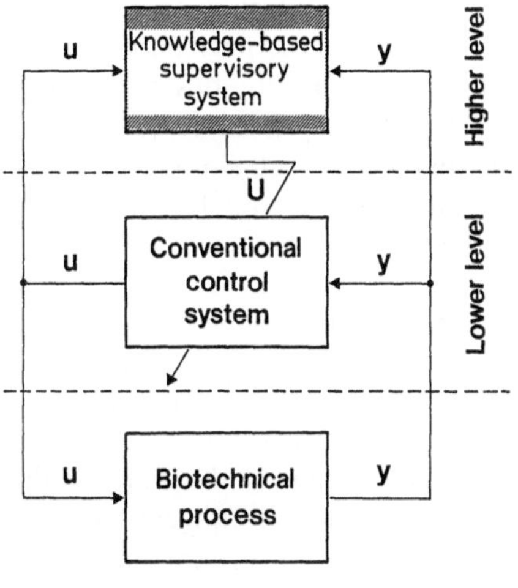

Fig. 2. Structure of a system for indirect (supervisory) expert control. The knowledge-based module is not involved in the low-level control loop; instead it supervises the lower control level by issuing high level commands

are included in almost all modern biotechnical equipment, either in the form of hardware controllers, or as computer-implemented algorithms. The higher level represents a knowledge-based superstructure to the conventional control part. In contrast to the direct scheme, the knowledge-based module, which is indeed a complete ES, is not explicitly involved in the low-level control (although exceptions are sometimes possible); it does not generate signals to the control plant, but helps the lower level to perform its job better. To this end, the higher level issues supervisory commands which tell the lower level when to do what [7]. As the techniques for design of low level controllers are relatively well established, the synthesis of the conventional part of the system is not likely to cause special difficulties. Thus, the success of the overall system development is dependent mainly on the design of the knowledge-based module using an appropriate ESDT.

The indirect expert control concept offers new possibilities for development of high-performance intelligent systems for the control of bioprocesses. This structure flexibly combines the advantages of the traditional approach with those of ES technology. It allows enhancement of the control system by the capability of intelligent decision-making based on informal interpretation of the complex behavior of the living system. If properly implemented, this structure will be capable of covering various control problems which usually remain outside the scope of conventional systems. The subsequent discussion will focus on systems for indirect expert control.

2.2 Functions of the Knowledge-Based Module of the Control System

The features of the ESDT necessary for building the knowledge-based module are correlated to the functions which this module is expected to perform in real-time. Although considerable differences are possible according to the application, these functions can be summarized into four main groups:

– Identification of the state of the cell population. This includes continuous on-line evaluation of the physiological state of the cell population, informal interpretation of cell behavior, prediction of future states, detection and diagnosis of expected (e.g. stage transfers) or unexpected (e.g. deviations from normal behavior) physiological phenomena, and others. Undoubtedly, this is the most important, difficult and advanced function of the higher system level [8, 9].

– Identification of the state of the process equipment. The major problem here is the detection and diagnosis of instrumental failures, such as troubles with sensors or actuators. It has been pointed out that this function must be explicitly separated from the previous one [10]. Although this may be not easy in case of biological plants, it will contribute to the clear modularization of the knowledge base of the ES.

– Supervision of the conventional control part. Identification of the state of the cell population and the process equipment are passive procedures creating declarative information on the plant. To become useful, this must be mapped into high level commands for supervision and synchronization of the work of the conventional control part. The major purpose is to achieve intelligent handling and control of the physiological phenomena during the process. Typical supervisory commands are activation/deactivation of control loops (changes of the control strategy), modification of control parameters, changes of setpoints, etc. Particular attention should be paid to the strategy-switching capability, which is needed to respond to structural transformations in biological systems [11, 12]. In the case of drastic physiological transfers, the knowledge-based part may need to react by dynamic replanning of the process. Such advanced capability cannot be achieved by the enumerated supervisory command; it requires additional internal segmentalization of the knowledge-based module, and establishment of hierarchical relationships between its parts [13].

Supervisions of the system by high level commands is necessary to handle detected instrumental failures. The purpose is to compensate as much as possible for any eventual damage, and enable the system to maintain safe operation under abnormal conditions, that is, to ensure robust fault-tolerant behavior.

A similar function of growing importance is the intelligent monitoring of the biotechnical plant. This is necessary because of the complexity of modern measurement equipment, which requires special supervision. There is probably no better recourse than to entrust this task to the knowledge-based module, thus providing coherent functioning of all the system components.

– Advanced communication with the user. The knowledge-based module should be able to represent all the information about the process, as well as to explain its decisions and activities in a convenient form for the user. Indeed, this function is mandatory for every ES, but the real-time constraints impose some more specific requirements.

2.3 Implementation Schemes

According to the method of coupling the knowledge-based part and the conventional control part, two types of implementation scheme are possible:

– Interface scheme. According to this, the knowledge-based part resides on a dedicated computer, interfaced to the conventional control part (Fig. 3). Both parts are software and hardware independent of each other. Typically, the higher level is purchased (as an ESDT), developed and added separately, after the conventional part has been set in operation.

Today, due to lack of other realistic alternatives, the interfaced scheme is the standard solution of the bioprocess control problem [4, 14, 15]. This is because the ESDTs available so far are large programs designed with the interfaced structure in mind, meant to run on stand-alone computers. Furthermore, modern bioreactors are equipped with computerized low-level controllers which perform some of the tasks of the conventional control module, and are easy to link to the ES.

Though popular, an interfaced architecture has several shortcomings: the software of both system parts is provided by different suppliers, which hampers integration; communication between computers may cause problems; the information storage and the man-machine interface of the system is redundant [10, 16].

– Embedded scheme. In this case, the knowledge-based module is embedded into the real-time control environment, co-existing with the conventional control

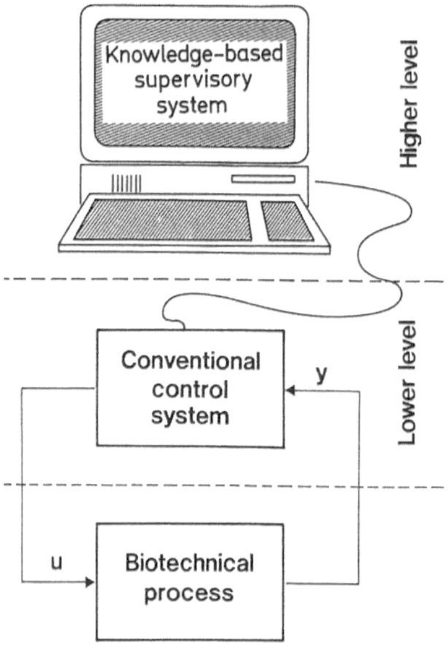

Fig. 3. Interfaced scheme. The knowledge-based supervisory module is implemented as an independent expert system on a separated computer interfaced to the lower level control system

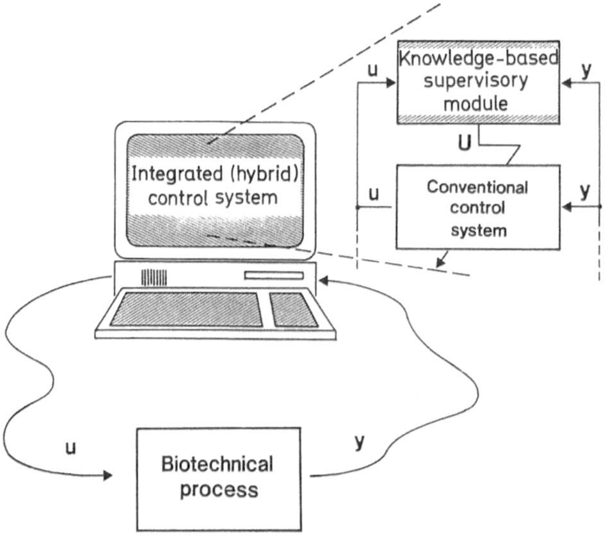

Fig. 4. Embedded scheme. The knowledge-based supervisory module coexists with the conventional control system on the same computer. The knowledge-based module is embedded into the real-time control environment

algorithms on the same computer (Fig. 4). The integration between both parts is much more efficient, and the problems of the interfaced scheme are overcome [17]. Although the embedded scheme is considered to be more advanced, its implementation is still difficult because the commercial ESDTs available today are not designed for such integration [18].

3 Requisite Features of Expert System Development Tools

Some of the features which an ESDT used to control bioprocesses should possess are common to all ESDTs designed for real-time applications, while others are more specific, and peculiar to bioprocesses. Below, the general features are described first.

3.1 General Features

3.1.1 Real-Time Capabilities

The real-time capabilities of the ESDT are critically important in process control. They have considerable impact on all other system features. For this reason ESDTs are classified into two groups: real-time and non real-time. In fact, the term "real-time" does not represent a single feature, but a complex of several characteristics:

– Fast response time. The speed of execution has always been subject to careful consideration in computer control systems. This is particularly important for ESs, which due to the processing of symbolic information, are known to be slow and clumsy. The speed of the conventional ES is estimated as two to three orders of magnitude lower than that required for real-time work [19].

The response time required in bioprocesses control is highly dependent on the particular application. Generally, the speed limitations are not so rigid as in the control of other technological processes, and will vary from few seconds (to handle fast phenomena, e.g. DO jumps, instrumental failures) to several hours (to interpret gradual physiological transfers). If the response time of an ESDT is not known, it can be practically estimated using some simple procedures as described in the literature [20].

Typically, high speed ESDTs are developed not in a symbolic language such as Lisp or Prolog, but in a fast conventional language such as C. Considerable reduction of the response time is achieved by using an ESDT which is based on compilation of the knowledge bases from a textual into a more efficient compressed form [3, 19]. However, as this procedure reduces the possibilities for on-line knowledge editing, and ESDT combining both compilation and interpretation of the knowledge base (called "incremental compiling" [21, 22]) appears to be the best solution.

– Activation of rules according to the time. This means that the system should be able to interpret instructions of the type

Run Rule 15 every 30 s

i.e. allow the attachment of a specific activation interval to every rule [23]. Furthermore, it is desirable to be able to change dynamically these time intervals from withing the rule conclusions, for example[1]

If (Process stage is State 2) and (DO is VERY LOW)
Then (Run Rule 15 every 5 s)

This can be useful in controlling bioprocesses because of their variable dynamics. Time-driven rule activation also provides temporal grouping of rules, which reduces the load of the interference mechanism; at any moment, only those rules whose time interval expires need to be processed.

– Acting within a time limit (guaranteed response time). Without such a mechanism the response time of the system is difficult to predict. It depends on the size of the knowledge base, the number of concurrently running tasks, and the availability of the data needed for the inference. If, for example, the data from the auto-analyzer are not yet available, but the decision cannot be postponed further, the system must be able to infer the best possible decision at the current moment [24]. To this end, the interference mechanism should be provided with a time-out mechanism which will interrupt the normal reasoning process when the permitted time-interval expires, and will force urgent, probably sub-optimal, decision making [18, 25].

– Cyclic continuous operation. Contrary to conventional ESs, which terminate after completing the decision, the ESs used in process control should be able to work continuously, in a never-terminating cycle [26]. Although this requirement seems to be simple, it is not always satisfied. A problem arises because many of the ESDTs are developed in symbolic language, usually Lisp. To prevent accumulative memory consumption during a long run, this language is provided with a so called "garbage collection" mechanism, which is automatically activated in unpredictable time instances. When running, the garbage collector interrupts for seconds, or even for minutes, all other operations, including the inference mechanism [22], which is absolutely inadmissible in a real-time environment.

In some new Lisp compilers the traditional "mark and sweep" garbage collection mechanism is replaced by a more advanced "incremental" type. An ESDT based on such a compiler would not have a problem with the interruption of continuous work.

There are several other characteristics, extensively discussed in the literature, which are needed to complete the real-time profile of the ESDT [18, 23, 24]. Any of them may be more or less important, dependent on the application. Unfortunately, most of the ESDTs available today do not support such features, and are not designed for real-time applications [19, 27]. There are, however, a few

[1] For simplicity, the exemplary rules are written in natural language. In real cases, they should conform to the ESDT syntax.

products which do provide some real-time capabilities [28], and it is on these that the attention of the developed should be entirely focused.

It is worth noting, that to a great extent the real-time capabilities of the ESDT are an inheritance from those of the operating system; if the operating system is not real-time, the ESDT will lack real-time features too. Therefore, tools designed for non-real time operating systems (e.g. DOS), are basically inappropriate for control applications. From this viewpoint, the most suitable ESDTs are those working in real-time UNIX (or UNIX-like) environments [29].

3.1.2 Temporal Reasoning

This is the capability of the ES to reason about discrete or continuous events and phenomena, accounting for the time dimension [30]. Temporal reasoning is closely related to the real-time features of the system. In order to emphasize the importance of temporal reasoning, it is discussed separately. Temporal reasoning has different aspects, some of which can be extremely useful in bioprocesses control:

– Reasoning about the history of continuous process variables [30]. The ES must have the ability to detect patterns, e.g. constraint violations, trends, or shapes, in specified pieces of historical data. This is illustrated by the template rule

If (Time-period Variable Descriptor)
Then (. . .)

where *Time-period* defines a period from the process history (often the most recent one) either explicitly, or in respect to a certain event in the past, *Variable* is a particular process variable, and *Descriptor* represents the expected pattern. The rule

If (SGR is NEGATIVE) and (Since entering Phase 3 RQ > 0.9)
Then (Deactivate the glucose feeding algorithm)

illustrates a simple form of reasoning in respect to the current value of the specific growth rate *SGR* and the recent history of the respiration quotient *RQ*. A slightly more complicated case is represented by the rule

If (The SGR has been DECREASING for more than 1 h)
Then (The process is entering the LATE GROWTH STAGE)

whose condition considers the recent trend of the SGR. There are many cases when the information from a biotechnical system can be adequately interpreted by accounting for the trends of variables. However, this is not always sufficient to analyze the behavior of bioprocesses in which the temporal shapes of the variables provide indispensable information about the underlying physiological phenomena. To achieve better control, these shapes must also be properly interpreted. This would be quite probable because the decisions of a human operator are based both on the current values and the recent historical profiles of the process variables. The rule

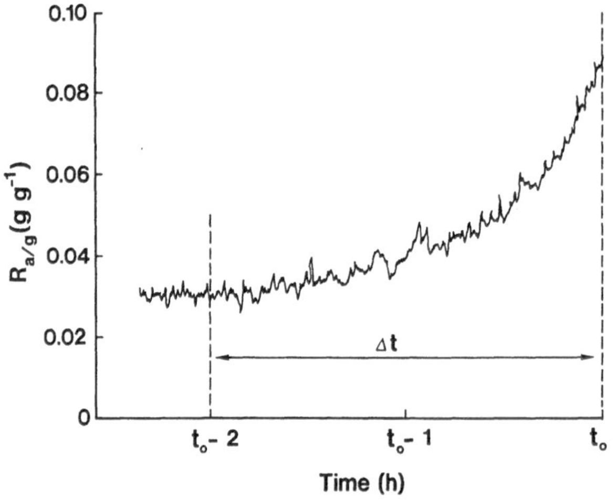

Fig. 5. Characteristic time profile of the $R_{a/g}$ indicating intensive excretion of acetic acid during cultivation of *Escherichia coli*

If *(During the last 2 h $R_{a/g}$ has INCREASED CONCAVELY)*
Then *(Report: Intensive excretion of acetic acid)*

represents such a case. It accounts for the shape of the variable $R_{a/g}$ (ratio of the ammonia feed rate to the glucose uptake rate), and will execute its conclusion if this shape matches the descriptor *"INCREASED CONCAVELY"* (Fig. 5). The shape descriptors may be different, e.g. *"PASSED OVER MAXIMUM"*, *"STOPPED INCREASING"*, *"STOPPED DECREASING"*, *"DECREASING MONOTONICALLY"*, *"OSCILLATING"*, possibly stored in a user-extendible library of shapes which the ES can recognize [10, 30, 31, 32, 33].

Another useful form of reasoning about the history of variables utilizes the data over the given time period for calculation of certain statistical markers, and decision making based on their values [16]:

If *(During the last 5 min the OD STANDARD DEVIATION is outside the interval [0.1, 2.0])*
Then *(Report: Problem with the OD sensor)*

The corresponding patterns are shown in Fig. 6.

– Reasoning about past events and phenomena. Except for continuous process variables, reasoning in respect to time can be applied to some events and phenomena which have been registered by the ES in the past, and whose occurrence may affect the decision in the current situation. The following example illustrates a simple case:

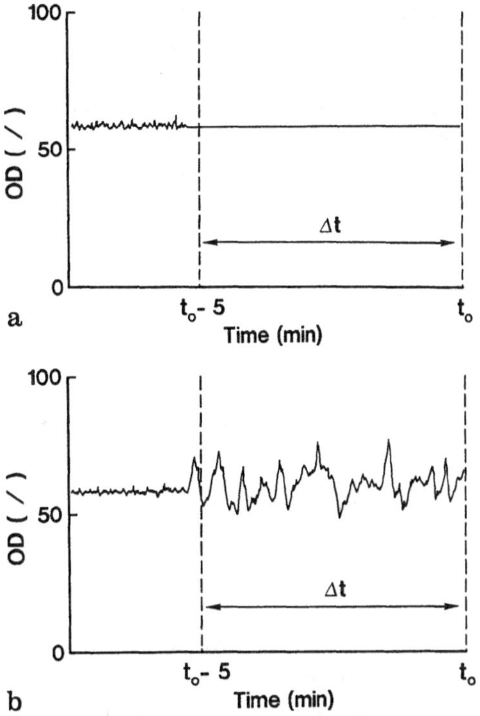

Fig. 6a, b. Two different patterns indicating failure of the *OD* sensor: **a** the normal noise level suddenly drops to zero; **b** the noise level increase abruptly

If (*The time after beginning of continuous feeding* $\geq$ *5 min*) *and* (*DO is HIGH*)
Then (*Increase the GFR by* ΔGFR)

where *GFR* is the glucose feed rate (see Fig. 7). To process this rule, the system should remember when the continuous glucose feeding began, i.e. to "time-stamp"

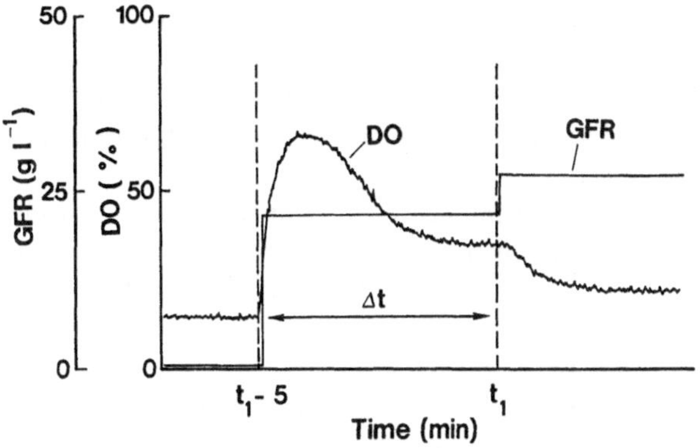

Fig. 7. Typical time profiles of the *DO* and the *GFR* upon transfer from batch to continuous glucose-limited cultivation. If the initial *GFR* is not sufficient, *DO* will remain high, which can be used to correct the *GFR*

this event. Such time-stamping of all detected events is another typical feature of the real-time ES, which always stores facts in the knowledge-bases together with the time of their appearance.

A slightly different form of temporal reasoning is represented by the introduction of lower and upper time bounds (permitted interval) for expected events. The system must then check whether or not the occurrence of the event fulfills these time constraints, for example

If *(Transfer Stage 1 → Stage 2 is outside the interval [20–26] h)*
Then *(Report: The transfer is time-incompatible)*

According to Lun and MacLeod [27], temporal reasoning mechanisms should support the construction of more abstract relationships, using time predicates such as BEFORE, AFTER, and DURING. This will facilitate reasoning about the order of the events in time.

Although temporal reasoning functions appear natural and useful, their realization raises many problems which have not yet been solved completely, even in specialized real-time ESs [1].

3.1.3 Integration with External Software Modules

Figure 2 shows that the knowledge-based part of the control system is not an isolated, independent module. To perform its function, it needs to exchange information with the external environment. This means that the ESDT must be provided with powerful communication mechanisms.

First, it should be integrated with the external software modules performing the conventional control functions. As it is virtually impossible to provide the ESDT originally with all algorithms needed in specific applications, the capability of linking with user-defined algorithms is an absolute requirement [23]. It has been reported that the power of the advanced shell G2, applied to bioprocess control, can be greatly enhanced by interfacing to user-defined algorithmic procedures for modeling, parameter identification, estimation of unmeasurable variables, prediction, signal processing, on-line optimization, statistical data-processing, and modeling of non-linear relationships by neural networks [14]. The integration should be understood either as incorporation of external software modules into the body of the ES, or as an establishment of communication channels between the ES and these modules.

Second, the ES should be able to communicate with hardware devices using, for example, serial lines. In other cases it may be necessary to retrieve or send data directly to the low-level process interface, e.g. AD/DAC and DI/DO. Since the communication with such devices is realized by software, the problem is again reduced to integration of the ES with the user's programs.

Most of today's ESs are not designed to communicate with conventional real-time software, but only with the operator through the keyboard [19]. This limitation is considered to be among the most serious reasons still preventing ES technology from being adopted by many industries [34, 35]. Some new products do, however, provide consistent integration capabilities using two basic mech-

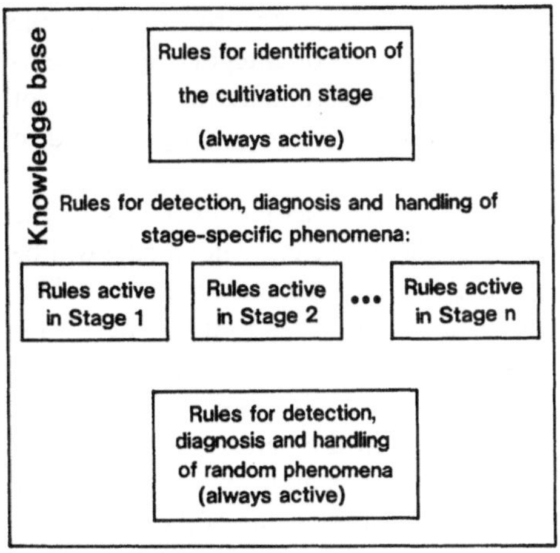

Fig. 8. Typical structurization of the knowledge base in the case of batch or fed-batch cultivation

anisms: fixed-protocol messages, which support communication between the ES and the external modules, and "software hooks", which are used for incorporation of the user's code directly into the ES. The last one is much more efficient, but unsafe because errors in the user's program may cause the ES to crash. Deep fusion at the level of knowledge- and data-bases, resulting in a so called "fully integrated" ES [16], is at present impossible.

3.1.4 *Mechanisms for Knowledge Structuring*

The knowledge of a bioprocess is not a homogeneous, plain bank of information. It is highly structured according to certain functional criteria, or in respect to time. For example, in the case of fed-batch cultivations, different chunks of knowledge are required in each of its stages (Fig. 8), i.e. at any moment, only a specific group(s) of rules should work. Application of rules which do not belong to the current context might be undesirable, even dangerous. Similar considerations hold true for continuous cultivations; however, the structuring will not have a chronological character, but the groups will correspond to possible process situations. To manipulate knowledge dynamically, the ESDT must provide mechanisms for activation or deactivation of a particular group(s) of rules. It is especially useful to do this from within the rule conclusions, for example:

If (RQ is LOW) and (SGR is LOW) and (SGR is DECREASING)
Then (Report: Transfer to Stage_2)
Then (Activate Rule_group_2)
Then (Deactivate Rule_group_1)

The explicit grouping of rules represents the simplest method of knowledge structuring. Although in many cases it will work successfully, high-performance ESDTs offer more sophisticated mechanisms for determining the currently active set of rules. They are usually based on some associative criteria, for example all rules which are related to a particular class of objects or category of problems [24]. The capabilities of driving the inference engine to narrow its scope of interest, and to use only part of the knowledge, are sometimes jointly referred to as "focus-of-attention" feature.

Another consistent way to structure knowledge is to assign priorities to the rules. The priority represents the importance of the particular rule; a rule will be checked only after all other rules with higher priority have been processed. This mechanism is useful because, by assigning different priorities to different groups of rules, one part of the knowledge base can be forced to work before another [36].

Apart from the logical clarity, there are two important advantages of knowledge structuring. First, there is a dramatic improvement in the performance speed of the system due to the reduced number of rules which require processing at any moment. Second, knowledge structuring simplifies the debugging of the knowledge base.

3.1.5 Handling of Various Categories and Levels of Knowledge

A fundamental feature of ESDTs is their flexibility in knowledge representation. This is crucial in the field of bioprocesses control, where various categories of knowledge are involved, ranging from superficial experiences to fundamental analytical models [37, 38]. Usually, the development of the control system is the final task of multi-phase research, including genetic, biochemical, microbial and behavioral study of organisms of interest. To take advantage of the information accumulated at each of these stages, the ESDT must provide rich capabilities for the representation and handling of diverse types of knowledge.

Although the ES approach is intuitively considered as heuristical, heuristics usually represent just a part of the knowledge available. The ES capabilities should not be limited only to representation of knowledge of this type. Knowledge of the bioprocess may be available in the form of analytical models (differential and difference equations, material and energy balances, kinetic models, stochiometric equations), qualitative models, or as fundamental information about cell genetics and metabolism. Such knowledge is represented in different forms, and the ES should be able to handle any of them equally well. This implies a simple rationale: when developing a new ES, the knowledge available should be analyzed first, the involved knowledge categories clarified, and based on this, the ESDT selected [39].

In addition to form, knowledge of the process differs also in depth. ESs are often criticized as being shallow models of their application domain, in the sense that they draw conclusions directly from the observed phenomena, without penetrating the underlying mechanisms. Recently, it has become evident that ES performance can be much improved by enrichment of this shallow knowledge by deeper knowledge, which provides consistent interpretation of the events, based on the underlying mechanisms, structures and dependencies [40, 41]. Often, deep knowledge is represented in the form of mathematical models. These can be very useful in solving tasks for the detection of sensor failures, deviations from expected behavior, estimation of inaccessible variables, forecasting, planning, hypotheses validation and knowledge-testing simulation [42]. From this viewpoint, ES technology should not be considered as a contradiction to modeling techniques; instead, the significance of the analytical models in ESs is expected to become more profound [43].

It would be incorrect to consider that deep knowledge can be represented only by mathematical models. It may exist in other forms, such as qualitative models [40, 44, 45], as linguistic models using English-like expressions [46], or even as set of rules. As our knowledge of the process represents a continuum, rather than clearly segregated layers, it is impossible to define a boundary between shallow and deep knowledge; knowledge considered as deep for one purpose, appears to be shallow for another. It is up to the system developer to decide the most appropriate depth of description in the particular application, but in any case the ES should not be the limiting factor in knowledge base development.

An important point to raise is that, in many cases, shallow knowledge is sufficient to solve various control problems. Let us consider the rule

If ($R_{a/g}$ *is HIGH*)
Then (*Reduce GFR by* Δ*GFR*)

which illustrates how to detect and how to prevent the excretion of acetic acid in a glucose-limited *Escherichia coli* cultivation [47]. This rule is a typical example of shallow knowledge. Nevertheless, it can work (and it really does!) quite satisfactorily. It represents, however, a "shortcut" through knowledge, which eliminates all intermediate steps, forming the causal link between the condition and the conclusion. The underlying physiological phenomena remain hidden, and the observed fact is mapped directly into the control decision. However, although very fast and memory-efficient, such rules reduce the capabilities of the ES. They treat the plant as a black-box, and restrict any possibility of explaining the observed phenomena. The eliminated intermediate conclusions will not be available in the knowledge base, and cannot be used by other rules in the system.

To describe the phenomena in more detail, the above rule can be expanded into a chain of rules. These will show that when the glucose feeding is high the cell oxidative capacity is exceeded, and cells start to excrete acetic acid, which results in a raise in the $R_{a/g}$ value. For more precise quantitative description of the phenomenon, a mathematical model can be added to the rules. Undoubtedly, the detailization may continue, involving more and more deep knowledge, down to the level of enzymes and pathways. The question is whether or not this is useful or efficient. Generally, excessive overloading of the ES with ballast knowledge is also harmful. As the control problem itself is usually formulated at a rather high level of abstraction, extremely detailed description can hardly be useful. Furthermore, the vast amount of knowledge will put a serious load on the computer, causing an unacceptable slow-down in the real-time environment. On the other hand, in some cases inclusion of deep knowledge may cause some heuristics to become redundant, resulting in a net reduction of the size of the knowledge base [44]. Therefore, the depth of the knowledge must be well balanced with the control purposes, the desired explanation capabilities, and the time and memory constraints.

3.1.6 Efficient Knowledge Debugging and Integrating

Building of a new knowledge base is an incremental and repetitive task. A few rules should first be defined and tested, and then a few more added. After every update, knowledge must be validated again. This is a mandatory step, since initially the knowledge will probably contain a couple of logical bugs and inaccuracies. To facilitate their localization and elimination, the ESDT should provide proper mechanisms, generally referred to as knowledge debuggers.

ESDTs usually include capabilities for off-line debugging of the knowledge base, such as the planting of breakpoints into the rules and step-wise tracing of the inference process. More advanced systems provide special mechanisms called knowledge-integrity checkers, which scan automatically the rule set for any logical inconsistencies, such as unreachable clauses, dead-end clauses, cyclic clauses, redundant rules, and subsuming rules [48].

Before real operation, the knowledge base should be tested also in realistic real-time conditions. This is usually achieved by some kind of simulation. The simplest way is the replacement of the sensor inputs by real cultivation data, collected in previous experiments. However, not all situations which may arise in practice can be checked in this way. As it is impossible to sacrifice a whole process to deliberately provoke the desired situation (sensor failures, nutrient limitation, contamination), this must be achieved by computer simulation. For such purposes the most advanced ESDTs are equipped with special simulators. This facility is considered to be one of the most powerful mechanisms for knowledge debugging [39].

Another important ESDT feature is the editing of old and integrating of new knowledge during on-line operation [49]. As the knowledge validation problem cannot be solved 100% off-line, errors are often discovered during the real run. These have to be resolved, without stoppage of the system, by on-line knowledge editing. Similarly, during cultivation, new phenomena or events may be observed. Instead of waiting until the end of the cultivation, it might be better to enrich the knowledge base by the new rule immediately. Similar conclusions can be drawn for editing and integrating other types of knowledge, e.g. analytical models.

3.1.7 Answering User's Questions

The complexity of bioprocesses makes the understanding of the accompanying phenomena by non-experienced operators difficult. Consequently, the activities of the ES will remain obscure for the novice, unless the logic of its operations is represented in a transparent form. This is achieved by an ES explanation facility, which allows the user to ask questions during operation; the ES will give answers based on the knowledge available and the current context. Consistent on-line explanation would contribute to the understanding of the process and the underlying control concepts. It would also increase the user's confidence in the ES, and may serve as a knowledge testing tool.

So far, the explanation facilities of ESDTs are rather restricted [50]. They are reduced to interpretation of a limited number of questions, typically "Why do you need this information?" and "How did you arrive at that decision?". These questions are easy to reply to because the answer is constructed in a straightforward manner, simply by looking one step up or down in the inference chain [22].

More sophisticated systems can answer questions such as "What are you doing now?", "What will happen if ...?", "What do you know so far about ...?", etc. Related features are the interactive explanation, and explanation at different levels of complexity according to the user's competence.

It has been shown that from an informational viewpoint the knowledge required for consistent explanation is more than that needed for the work of the ES [51]. Today's ESs try to explain their decisions and actions, i.e. their knowledge, by means of that knowledge itself. Generally, for intelligent explanation, deeper knowledge than that used in problem-solving, is required. However, enrichment of the knowledge base by such supplemental information just for explanation purposes will always slow down the inference process. A possible solution is the development of a separate knowledge base especially for explanations. In many ESDTs, this feature, though not supported, is emulated in a purely mechanistic way. It allows the developer to attach explanatory texts to the rules, which will appear on the screen as answers to a user's questions under particular circumstances. This mechanism is primitive, but it can be useful in many cases.

It is expected that improvement of the explanation facilities will be possible by narrowing the ES application area [50]. For example, in the future, ESDTs specialized for bioprocesses control that answer questions such as "How has the current physiological state been reached?", "What is your prediction about the next state?", "Explain the growth profile", "Why is OUR decreasing?", might be designed.

3.1.8 Advanced User Interface with Graphical Capabilities

This feature is relatively well developed today. The main difficulties ensue from the necessity to combine simplicity of the man-machine interface with the advanced performance of the ES. In bioprocess control, it is extremely important to maintain simplicity because potential users tend to lack specialized computer skills. Fortunately, modern software techniques for graphical representation, menu-driven dialogs, icon definitions, simplified data input using a mouse, separation of the developer interface from the end-user interface, and others, provide a fairly good balance between simplicity and sophistication.

A major role in the user interface is played by the graphical capabilities of the ESDT. Since in bioprocesses the main part of the information is represented by continuous variables, graphical features are of utmost importance [52]. Today, these are supported by almost all advanced ESDTs, though some companies are selling the graphics packages as a separate product. Apart from the representation of process variables, modern systems use graphics for the development and representation of schemes, tables and diagrams. These help in providing more realistic visualization of the situation in the control plant [24], or in representing the rule base as an easy-to-understand network [36, 37].

3.1.9 Other Features

ES technology is advancing rapidly, and modern tools are enhanced by additional features. Among them, the capability of *object-oriented representation*, which is regarded as being very natural for biotechnological applications, deserves special attention [53, 54]. However, this feature is provided in different products at different levels. Some ESDTs, which originally lacked object-oriented representation, have recently been modified to include it. The resulting "sewn on" capabilities are usually restricted, because the internal ESDT structure, created initially without an object-oriented approach in mind, cannot be completely reorganized to handle the new paradigm [36].

The *neural network* approach, which has recently proven useful in modeling systems of unknown structure, including bioprocesses [14, 55], is another issue concerning ESDTs. It has been shown that ESs can incorporate neural networks to enhance their problem-solving power [56]. This capability should be also considered when selecting an ESDT.

Learning is a feature which allows progressive improvement of the system performance. It is inherent to the neural network approach, but in the ES field is still a subject of theoretical research [1]. Nevertheless, some positive results of the application of learning algorithms in bioprocess control have already been published [8].

ES progress in the field of the man-machine interface is being stimulated both by developments in theory and improvements in hardware. Some ESDTs are already equipped with *natural language processors*; further enhancements in the areas of *video* and *audio* "multimedia" capabilities are also expected [57].

There are many other features of ESDTs described in the specialized literature, which could be more or less important in bioprocess control. It is worth noting that together with their useful capabilities, much attention is often given to some "fancy" features. The serious designer should always take care to differentiate between really feasible issues, and such unnecessary elaborations [58, 59].

3.2 Specific Features

In addition to the general features discussed above, there are some issues related more specifically to the control of bioprocesses. They reflect the unique characteristics of biological systems, which differ from those of conventional control plants, and influence the selection of the ESDT.

3.2.1 Limited "Width" and Enhanced "Depth"

Unlike in other fields, in bioprocesses control, the set of the on-line measured variables is quite limited. To minimize the risk of contamination, the number of sensors in large-scale systems is even fewer than in small-scale laboratory bioreactors. This implies that the capabilities of scanning tens of thousands of inputs and maintaining huge data bases, provided by some commercial ESDTs [28, 60, 61], are not necessary. Also, it is not expected that the size of the ES rule base will be large. The number of rules will probably range between several dozen and a few hundred [42, 62, 63]. Thus, it is not likely that ESs for the control of bioprocesses will grow too much in width.

The reduced size of the knowledge base does not, however, mean simplicity. To interpret intricate and vague situations, the form of rules and the structure of the knowledge base will be complex. Consequently, a compact and flexible ESDT, providing a rich set of functional capabilities (referred to as the "depht" of the ES) will best suit the problem of interest.

3.2.2 Handling of Uncertain, Incomplete, and Fuzzy Information

The problem of the amount and quality of the information available on-line in biotechnical systems is probably more serious than in any other control plant. Many of the physiological phenomena are poorly understood, large regions of the physiological state space remain unknown, while the information about others is not crisp and quantitative, but fuzzy, with a qualitative character. In addition, due to the lack of sensors for biochemical variables, it is impossible to supply the control system with all the information necessary on-line (note that even an adequate cell concentration sensor is not yet available!).

This imposes a serious requirement on an ESDT used for bioprocesses control: it must be able to handle and utilize such uncertain, incomplete and fuzzy information, to make decisions in conditions of concurrent solutions, and to act adequately in "unknown" situations. Indeed, the problem is related to one of the main tasks of the higher level of the control system (Fig. 2), which must resolve process uncertainties, creating a simple deterministic environment for the functioning of the lower system level. Such capabilities are usually implemented by mechanisms for handling fuzzy sets, qualitative knowledge, certainty factors, and confidence levels, which are provided by some of the ESDTs available today.

The processing of rules of different types (crisp or fuzzy) requires dynamic alteration of the method for calculating the certainties of their conclusions from the certainties of the facts in the condition. The ESDT should provide such flexibility, possibly by assigning the preferred method to every rule as an argument. In the author's experience, such a feature is of fundamental importance in the handling of complex and uncertain phenomena [8, 17], but it is not supported by commercial ESDTs.

3.2.3 Orientation Towards the Processing of Continuous Variables

Almost 100% of the information available on-line is in the form of continuous process variables with complex dynamic behavior. Because of the lack of sensors, the control system must "squeeze" as much information as possible from the available variables and their histories. Therefore, capabilities of advanced signal processing, both deterministic and stochastic, are highly desirable.

3.2.4 Capabilities of Intelligent Monitoring

Because of the complicated dynamics of bioprocesses and the growing complexity of recent measurement equipment, intelligent supervision is needed not only for the control procedures, but also for the measurement tasks. According to the situation, various modifications of the measurement procedures might be required. Examples are a change of the sampling intervals according to the dynamics of the plant, activation/deactivation of the measurement or estimation of variables valid only in particular time-intervals, periodical on-line sensor calibration, range switching, supervision of the work of advanced analytical devices (mass spectrometers, flow injection- and other analyzers, auto samplers, etc.), and control of some specific measurement procedures which require test actions. A simple example is the on-line calibration of exhaust gas analyzers. This procedure causes a jump in the CO_2 and O_2 concentrations, and disturbs all related variables (OUR, CER, RQ, etc.) which might be used for control purposes. To prevent incorrect decisions and actions, the ES should suspend the updating of these variables during the calibration period.

The ESDT must be flexible enough to accommodate knowledge for supervision of the measurement procedures. Enrichment of the knowledge base with such rules will provide the measurement with a kind of intelligence too, and will contribute to the synchronization and coordination of the ES activities.

3.2.5 Availability of Common "Bioprocess" Knowledge Bases

General ESDTs come with empty knowledge bases which must be filled by the knowledge engineer, or by the user himself. The knowledge bases of specialized ESDTs might be originally loaded with a certain amount of fundamental information, valid in most of the applications in particular field. It is believed that such organization will be useful in bioprocess control [14]. Some basic parts of the knowledge of biological systems concerning metabolism, population behavior, or other aspects of bioengineering, can be permanently fixed into the ESDT. In many cases this would shorten the development of the control system, and make the knowledge of the plant more comprehensive. Unfortunately, such a feature is not yet available. In future, it might be provided by an ESDT designed especially for the control of bioprocesses.

3.2.6 Low Prototyping Time

Modern biotechnology is an area of rapid technological evolution, in which the development of a new process, or the improvement of an old one, does not usually take long. It would be impractical if the effort required for building the control system were too time consuming, thus delaying the application period. Therefore, the time needed for ES development is another criterion which should be carefully considered. A terse and flexible, yet powerful ESDT is favored, whose application requires minimal study and programing effort. This is especially important for biotechnology teams, which normally do not include software specialists.

From a software viewpoint, there are several ways to create an ES. It can be made completely by oneself using a proper programing language, or the system can be constructed from an ESDT. According to the complexity of the task, the time for development of an ES from an ESDT usually varies from three months to about one year [58], which is generally acceptable for biotechnology applications. However, the time required for building a system using conventional language is estimated to be several times longer than using an ESDT [21]. The conclusion is straightforward: low prototyping time can be achieved only by using an ESDT. Indeed, this is the common practice in all ES application fields, with over 90% of the systems being based on ESDTs [58].

3.2.7 Low Price

Most of today's biotechnology laboratories are equipped with a number of bioreactors, often used in parallel for development of different processes. Depending on the circumstances, several of these reactors may need monitoring and control by an advanced computer system. Practically, it is possible to use a single ES for control of more than one process, but as strain characteristics, cultivation conditions, logic of operation and control objectives might vary tremendously, such integration would be artificial, resulting in a series of problems. In such a case, better solution is to provide each bioreactor with an independent control system. This is, however, possible only if the corresponding hardware and software are inexpensive.

Fortunately, the characteristics of the bioprocesses allow the application of non-expensive hardware, such as 32-bit PCs. As almost all ESDTs are available in PC versions [58], there is a common trend today towards the application of PCs in expert technology. However, the prices of the real-time ESDTs themselves are prohibitively high, ranging from several thousand dollars for systems with moderate capabilities, to above 50000 dollars for top-class ESDTs. Undoubtedly, the software price is another serious restriction, still preventing ES control technology from becoming daily biotechnological practice.

4 Conclusions

Today, there is no single ESDT which combines all the features discussed above, and since such a perfect tool may not be available soon, the developer should select the most convenient product from those currently on the marked. The most appropriate are ESDTs with build-in real-time capabilities, such as NEXPERT OBJECT, ART, KEE, and particularly those designed for process monitoring and control, e.g. ESCORT, R*TIME, PICON, MUSE, COMDALE and G2 (the last being considered to be the most advanced one [16, 28, 64]). Due to their flexibility, these ESDTs can be customized to a particular application, though such adaption will not satisfy all specific requirements. Nevertheless, the resulting ES will be capable of covering a large portion of the problems arising in the control of bioprocesses.

Certainly, the ideal ESDT in this field would be far more specialized. The lack of such a product from the market extends the time and effort needed for the creation of control systems considerably. Since large-scale use of ES technology in the area of bioprocess control seems inevitable, the development and commercialization of such an ESDT would be an important contribution to the biotechnology community.

5 References

1. Verbruggen HB, Åström KJ (1989) Proceed IFAC Workshop on AI in Real-Time Control, Shenyang, PRC, p 1
2. Halme A, Visala A (eds) (1989) Proceed International Workshop on Expert Systems in Biotechnology, Helsinki, Finland
3. Shirley RS, (1987) IEEE Control Systems Magazine 7: 11
4. Cooney CL, O'Connor GM, Sanches-Riera F (1988) Proceed Int Biotechnol Symp, Paris, p 563
5. Cardello RJ, San K (1989) Proc. American Control Conference, Pittsburgh, Pennsylvania, 2411
6. Konstantinov KB, Yoshida T (1992) Biotechnol Bioeng 39: 479
7. Smuts WB, MacLeod IM (1989) Proceed IFAC Workshop on AI in Real-Time Control, Shenyang, PRC, p 31
8. Konstantinov KB, Yoshida T (1989) Biotechnol Bioeng 33: 1145
9. Halme A (1988) Proceed IFAC Symp on Computer Applications in Fermentation Technology, Cambridge, UK, p 159
10. Lukas MB, Oyen RA, Keyes MA, Kaya A (1989) Proceed IFAC Workshop on AI in Real-Time Control, Shenyang, PRC, p 79
11. Konstantinov KB, Yoshida T (1990) J Ferm Bioeng 70: 48
12. Bakay A, Madaraasz L, Hinsenkamp A, Papp Z, Dobrowiecki T (1990) Preprints 11th IFAC World Congress, Tallin, USSR, 7: 264
13. Saridis GN (1989) Atomatica 25: 461
14. Aynsley M, Peel D, Morris A (1989) Proceed American Control Conference, Pittsburgh, Pennsylvania, p 2239
15. Aynslay M, Hofland AG, Montague GA, Peel D, Morris AJ (1990) Proc American Control Conference, San Diego, California, p 1992
16. Arzen KE (1989) Proceed American Control Conference, Pittsburgh, Pennsylvania, p 2233
17. Konstantinov KB, Yoshida T (1991) Proceed IFAC Workshop on Computer Software Structures Integrating AI/KB Systems in Process Control, Bergen, Norway, p 198
18. Beck T, Lauber J (1990) Preprints 11th IFAC World Congress, Tallin, USSR, 7: 158
19. Laffey TJ, Cox PA, Schmidt JL, Kao SM, Read JY (1988) AI Magazine 9: 27
20. Press L (1989) IEEE Expert 4 (Spring), p 37
21. Gevarter, WB (1987) Computer, 20 (May), p 24
22. Winston PH, Horn BK (1989) Lisp, Addison-Wesley, Massachusetts
23. Laffey TJ (1991) Byte (January), p 259
24. Moore R, Rosenof H, Stanley G (1990) Preprints Of 11th IFAC World Congress, Tallin, USSR, 7: 234
25. Wright ML, Green MW, Fiegl G, Cross P (1986) IEEE Software (March), p 16.
26. Ennis RL, Griesmer JH, Hong SJ, Karnaugh M, Kastner JK, Klein DA, Milliken KR, Schor MI, Van Woerkom HM (1986) IBM J Res Develop 30 (January), p 14
27. Lun V, MacLeod IM (1989) Proceed IFAC Workshop on AI in Real-Time Control, Shenyang, PRC, p 25
28. Coyle F (1990) IEEE Expert 5 (October), p 12
29. James JR, Suski GJ (1988) Proceed IEEE Conf on Decision and Control, Austin, Texas, p 580
30. Konstantinov KB, Yoshida T (1992) AIChE J. 38: 1703
31. De Feyter AR (1989) Proceed IFAC Conf Advanced information processing in Automatic Control, Nancy, France, p 231
32. Konstantinov KB, Yoshida T (1991) IEEE Trans Syst, Man, Cybern SMC-21: 908
33. Konstantinov KB, Yoshida T (1992) Proceed IFAC Symp on On-Line Fault Detection and Supervision in the Chemical Process Industries, Newark, Delaware, p 93
34. Rao M, Cruz R, Yang T, Jiang T, Wang S, Kim I (1989) Proc American Control Conference, Pittsburgh, Pennsylvania, p 2418

35. Haase VH (1990) Preprints 11th IFAC World Congress, Tallin, USSR, 1: 141
36. Payne EC, McArthur RC (1990) Developing expert systems, Wiley, New York
37. Aarts, RJ, Suviranta A, Rauman-Aalto P, Linko P (1989) Food Biotechnology 4: 301
38. Montellano R, Bernier M, Chery A, Farza M (1990) Preprints 11th IFAC World Congress, Tallin, USSR, 7: 54
39. Johannsen G, Alty JL (1991) Automatica 27: 97
40. Kuipers BJ (1986) Artificial Intelligence 29: 289
41. Price C, Lee M (1988) Proceed 12th IMACS World Congress, Paris, p 85
42. Lübbert A, Hitzman B, Kracke-Helm H, Schügerl K (1988) Proceed IFAC Symp on Computer Applications in Fermentation Technology, Cambridge, UK, p 297
43. Voss H (1988) Proceed IFAC Workshop on AI in Real-Time Control, Clyne Castle, UK, p 1
44. Aarts RJ (1989) Proceed International Workshop on Expert Systems in Biotechnology, Helsinki, Finland, p 108
45. Beaumont SF, Leyval L, Gentil S (1989) Proceed IFAC Conf Advanced Information Processing in Automatic Control, Nancy, France, p 181
46. Smith JM, Stutely R (1988) SGML: The users guide, Ellis Horwood, Chichester
47. Konstantinov KB, Nishio N, Seki T, Yoshida T (1991) J Ferment Bioeng 71: 350
48. Cragun BJ, Stendel HJ (1987) Int J Man-Mach Stud 26: 633
49. Walker A (1986) IBM J Res Develop 30 (January), p 2
50. Wick MR, Slagle JR (1989) IEEE Expert 4 (Spring), p 26
51. Chandrasekaran B, Mittal S (1983) Int J Man-Machine Studies (November), p 425
52. Stephanopoulos G (1987) Chemical Engineering Progress 83: 44
53. Stephanopoulos G, Stephanopoulos G (1986) Trends in Biotechnology 4: 241
54. Linko S, Aarts R (1990) Abstracts of the European Congress in Biotechnology, Copenhagen, p 326
55. Thibault J, Van Breusegem V, Chwruy A (1990) Biotechnol Bioeng 36: 1041
56. Rich E (1990) IEEE Expert 5: 5
57. Bielawski L, Lewland R (1991) Intelligent System Design: Integrating Hypermedia and Expert Systems Technologies, Wiley. New York
58. San Giovanni J, Romans H (1987) Chem Eng Progress 83: 52
59. Bristol EH (1987) Proceed IEEE Conf on Decision and Control, Los Angeles, p 1948
60. Moore RL (1985) Control Engineering (April), p 118
61. Satch PA, Paterson AM, Turner MH (1986) Expert Systems 3: 22
62. Chen Q, Wang S, Wang J (1988) Proceed IFAC Symp on Computer Applications in Fermentation Technology, Cambridge, UK, p 253
63. Xu E, Xu G, Zhang S (1989) Proceed IFAC Workshop on AI in Real-Time Control, Shenyang, PRC, p 97
64. Eldeib HK (1989) Proc American Control Conference, Pittsburgh, Pennsylvania, p 2637

Author Index Volumes 1–48

Subject Index